Experimental Design for the Life Sciences

Experimental Design for the Life Sciences

FOURTH EDITION

Graeme D. Ruxton

University of St Andrews

Nick Colegrave

University of Edinburgh

OXFORD
UNIVERSITY PRESS

Great Clarendon Street, Oxford, OX2 6DP,
United Kingdom

Oxford University Press is a department of the University of Oxford.
It furthers the University's objective of excellence in research, scholarship,
and education by publishing worldwide. Oxford is a registered trade mark of
Oxford University Press in the UK and in certain other countries

© Graeme D. Ruxton and Nick Colegrave 2016

The moral rights of the authors have been asserted

First edition 2003
Second edition 2006
Third edition 2011

Published in the United States of America by Oxford University Press
198 Madison Avenue, New York, NY 10016, United States of America

British Library Cataloguing in Publication Data
Data available

Library of Congress Control Number: 2016933213

ISBN 978-0-19-871735-5

Printed in Great Britain by
CPI Group (UK) Ltd, Croydon CR0 4YY

To Katherine, Amelie, Isla, and William

Preface

Why you should read this book

A little effort (this is a short book) will help you design better experiments, and designing better experiments will make your life richer and happier and possibly easier! First of all, collecting data to test theories is the very essence of science—no matter what sort of scientist you aim to be, you really need to be able to design good experiments yourself and/or evaluate other scientists' experiments. So don't fool yourself by pretending experimental design doesn't matter to you. We are confident that whether you have never designed an experiment before or whether you have been doing it for some time, there are ideas in this book that will improve your confidence and ability to design and evaluate data collection. Don't fool yourself into thinking that experimental design is an abstruse exercise in applied maths that is beyond you. Flick through this book, there is barely an equation, and we have spent four editions honing our explanations to make them as clear as possible—without a shadow of doubt, if you read this book, then you will understand it and you will be a better scientist for it. At the opposite extreme, be wary of thinking that experimental design is so simple that this book isn't worth your time. If that were true then the world would not be littered with student lab reports with low grades because the student made basic mistakes in experimental design, and manuscripts by professional scientists that will never become published scientific papers for the same reason. If you read this book then for the rest of your scientific life you will design experiments that give clearer answers to more interesting questions, using less of your time, fewer resources, and simpler statistics. Not only that, but you will get more from every scientific paper you read and every scientific talk you listen to. The upshot of these benefits is that you will be a better scientist and a happier person. Given all that, the question really is why on Earth should you not read this book?

Why not to read this book

If you want a book on statistics this is not it. We do explore links between experimental design and statistical analysis throughout the book, but this book has very little to teach you about formal statistics. If you want an encyclopaedic tome covering every possible experimental design, then we cannot offer that either: that would be a book at least ten times the size. What we do offer is to give you an understanding of the underlying philosophy common to all experimental designs, and a clear introduction to terminology—these will help you to read books and papers that offer complex designs, but here we only look in depth at relatively simple designs. But 'simple' does not mean 'toy'—the latest issues of the very best scientific journals have articles based on experiments no more complicated that the ones we cover in this book, and most scientists will go through their whole career without having to delve into more complicated designs.

Why we felt the need to produce yet another edition of the book

In truth, some typos aside, there is nothing that was downright wrong in any of our previous editions. With each new edition we have been responsive to feedback from readers and from our own development as scientists to come up with improvements in how we communicate the basic ideas of experimental design. This revision has been much more sweeping that the previous two. We have extensively reorganized and rewritten so as to emphasize linkages between different aspects of experimental design more effectively. Thus we think this book should function both to give you quick answers when you have a very focused question and to give you a solid overview of the overarching concepts of experimental design. From that perspective, we hope that you read the book at least once from cover to cover, and then keep it to dip into quickly as and when required.

Learning features

Key definitions

There is a lot of jargon associated with experimental design and statistical analysis. We have not tried to avoid this. Indeed, we have deliberately tried to introduce you to as much of the jargon as we can. By being exposed to this jargon, we hope that it will become second nature to you to use—and understand—it. This should also make reading more advanced texts and scientific papers on experimental design less daunting. However, to make negotiating the minefield of jargon more straightforward, we have increased the number of definitions of key terms provided and increased the level of detail in all definitions. Each key word or phrase is emboldened at the point where we first use it, and is given a clear definition in a box nearby.

Reverse causation is mistakenly concluding that variable *A* influences variable *B* when actually it is *B* that influences *A*.

Statistics boxes

To emphasize the important link between good experimental design and statistics we now include a number of statistical boxes. These boxes should help you to see how thinking about design helps you think about statistics and vice versa. We have added the boxes at points where we think that keeping the statistics in mind is particularly useful and include pointers in the main text to each box.

Reverse causation

The second problem of correlational studies is **reverse causation**. This can we see a relationship between factors *A* and *B*. It means mistakenly assu tor *A* influences factor *B*, when in fact it is change in *B* that drives chan

For example, imagine a survey shows that those who consider thems use recreational drugs also consider themselves to have financial wo tempting to conclude that a drug habit is likely to cause financial probl causation explanation is that people who have financial problems than average to turn to drugs (perhaps as a way to temporarily escap worries). In this case, we'd consider the first explanation to be the more reverse causation explanation is at least plausible. Indeed, both mechanis operating simultaneously.

STATISTICS BOX 3.1 Generalizing from a study

An important issue in the int erpretation of a study is how widely applicable the re- sults of the study are likely t o be. That is, how safely can you generalize from the particular set of subjects use d in your study to the wider world? Imagine that we measured representative sa mples of UK adult males and females and found that males in our sample were o average 5 cm taller. Neither we nor anyone else are particularly interested in specific individuals, rather we are interested in what we can conclude mo widely from study of the sample. This becomes easy if we about the population from which our sample was drawn. Our aim was to cre sample of UK adults. If we have done this well, then we should

Self-test questions

The more you think about experimental design, the easier
designing robust experiments becomes. To get you thinking
while you are reading this book, we include a number of self-test
questions in every chapter. Often there will not be a clear right
or wrong answer to a question, but suggested answers to all
questions can be found at the back of the book.

Take-home messages

To help to consolidate your thinking, we end most sections with
a take-home message. By including these throughout the text
we hope to give you an immediate opportunity to evaluate your
understanding of a section before moving on.

Boxes and online supplementary material

We have aimed to produce a book that can be read from cover
to cover, and so have tried to keep unnecessary details out
of the main text. However, in areas where we feel that more
details or more examples will lead to a fuller understanding of a
concept we have included supplementary boxes. Where there is
(to our mind) really interesting material that will not be strictly
relevant to all readers, we have not included that material in the
book itself, but lodged it on an associated website (www.oxford-
textbooks.co.uk/orc/ruxton4e/). We point to this supplementary material at appropriate points throughout the book.

Chapter outlines

Most chapters begin with an outline of the main points covered by that chapter. These
outlines should help you to prepare for what is ahead. The outlines will also provide an
easy way for you to dip in and out of the book during subsequent readings.

Chapter summaries

Every chapter finishes with a summary of the most important points covered in that
chapter. By reading through this list you should be able to reassure yourself that you
have got everything out of the chapter that you can, or go back and read sections
again that are not yet clear in your mind.

Flow chart

The exact process of designing an experiment will vary considerably between studies. Nevertheless, there are key stages in the design process that will apply to most, if not all, studies. The flow chart at the end of this book is intended to summarize and guide you through the main stages of designing an experiment. We have indicated at each point in the chart the sections of the book that are most relevant.

Acknowledgements from the fourth edition

In the 15 years we have been involved with this book, we have had explicit feedback from countless students and colleagues; and have had our thinking on experimental design sharpened by countless others. We don't attempt even a partial list; but if you spot your influence anywhere in this work be assured of our profound thanks. Even if you can't spot your direct influence be aware that we are both really grateful for everyone who has forced us to think harder about experimental design and how to explain our ideas. We are grateful for the support we have had from editorial staff at OUP (most recently Jessica White) who have always been supportive when required without ever feeling heavy-handed; the production staff (most recently Tomas Furby) have always been a model of quiet competence. Julian Thomas is a copy editor who combines thoroughness with a collegiate, open, and interactive approach. We have both been privileged to work at institutions that have fostered the academic freedom that has allowed us to devote so much time to this endeavour. We have both been blessed by friends and family that have been tolerant of the demands of this book but always there to provide compelling reasons not to let it swallow our lives entirely.

Contents

7 The simplest type of experimental design: completely randomized, single-factor   101

8 Experiments with several factors (factorial designs)   109

Why you should care about design

- We begin this chapter by explaining the benefits of learning how to design good scientific studies (Section 1.1), and the potential costs of failing to do so (Section 1.2).

- We then address the relationship between design of a study and statistical analysis of the data collected in the study (Section 1.3).

- Much of good experimental design comes down to dealing with the challenges of random variation and confounding factors, so we next introduce these concepts (Section 1.4).

- In any study, we generally measure only a sample from the wider population that we are interested in. We discuss how life scientists refer to the 'things' in a sample as an introduction to how we deal with the specialist terminology of experimental design throughout the book (Section 1.5).

1.1 Why experiments need to be designed

When some life scientists see the phrase 'experimental design', it can either ease them gently into a deep sleep or cause them to run away screaming. For many, experimental design conjures up unhappy memories of mathematics or statistics lessons, and is generally thought of as something difficult that should be left to statisticians. Wrong on both counts! Designing simple but effective experiments doesn't require difficult maths. Instead, experimental design is more about common sense, biological insight, and careful planning. Having said that, it does require a certain type of common sense, and there are some basic rules. In this book, we hope to steer you relatively painlessly towards thinking more effectively about designing experiments.

So why are many life scientists so averse to thinking about design? Part of the reason is probably that it is easy to think that time spent designing experiments would be better spent actually doing experiments. After all, the argument goes, we are biologists so let's concentrate on the biology and leave the statisticians to worry about the design and analysis. This attitude has given rise to two myths that you can hear from the greenest student or the dustiest professor.

> **Myth 1:** It does not matter how you collect your data, there will always be a statistical 'fix' that will allow you to analyse it.

It would be wonderful if this was true, but it is not. There are a large number of statistical tests out there, and this can lead to the false impression that there must be one for every situation. However, all statistical tests make assumptions about your data that must be met before the test can be meaningfully applied. Some of these assumptions are very specific to the particular test. If you cannot meet these, there may be a substitute test that assumes different characteristics of your data. But you may find that this alternative test only allows you to use your data to answer a different scientific question to the one that you originally asked. This alternative question will almost certainly be less interesting than your original (otherwise why weren't you asking it in the first place?). Further, there are some basic assumptions that apply to many of the most commonly used statistical tests, and you ignore these at your peril. For instance, statistical tests generally assume that your data consist of what statisticians refer to as **independent data points** (more on this in Chapter 5). If your data don't meet this criterion then there is much less that statistics can do to help you. Careful design will allow you to avoid this fate. It's also generally true that well-designed experiments require simpler statistical methods to analyse them than less-well-designed experiments. So time spent carefully designing your experiment might save you a lot of time later in the study trying to master more complex statistical techniques than you really needed.

The group of experimental subjects that we use in our experiment (called the sample) needs to be representative of the wider set of individuals in which we are interested (called the population). One key way to achieve this is to ensure that each individual selected is not linked in some way to another individual in the sample, i.e. they are independent. For example, if we were surveying human food preferences, then gathering food-preference data from five members of the same family would not produce five independent data points. We would expect that members of the same family are more likely to share food preferences than two unconnected individuals, since family members often have a long history of eating together. Similarly, gathering data from the same person on five separate occasions certainly does not provide five independent data points, since a person's preference on one occasion is likely to be a good guide to their preferences a little while later.

> **Myth 2:** If you collect lots of data something interesting will come out, and you'll be able to detect even very subtle effects.

It is always reassuring to have a notebook full of data. If nothing else, it will convince your supervisor that you have been working hard. However, quantity of data is really no substitute for quality. A small quantity of carefully collected data, which can be easily analysed with powerful statistics, has a good chance of allowing you to detect interesting biological effects. In contrast, no matter how much data you have collected, if it is

---

Independent data points come from unconnected individuals. If the measured value from one individual gives no clue as to which of the possible values the measurement of another individual will produce, then the two measurements are independent.

Q 1.1 If we wanted to measure the prevalence of both left-handedness and religious practices among prison inmates, what population would we sample from?

Q 1.2 If we find that two people in our sample have been sharing a prison cell for the last 12 months, will data from them be independent?

of poor quality, it will be unlikely to shed much light on anything. More painfully, it will probably have taken far longer and more resources to collect than a smaller sample of good data.

➡ Designing effective experiments needs thinking about biology more than it does mathematical calculations. Careful experimental design at the outset can save a lot of sweat and tears when it comes to analysing your data.

1.2 The costs of poor design

1.2.1 Time and money

Any experiment that is not designed in an effective fashion will at best provide limited returns on the effort and resources invested, and at worst will provide no returns at all. It is obvious that if you are unable to find a way to analyse your data, or the data that you have collected does not enable you to answer your question, then you have wasted your time and also any materials. However, even if you don't make mistakes of this magnitude, there are other ways that a poorly designed experiment might be less efficient. It is a common mistake to assume that an experiment should be as big as possible; but if you collect more data than you actually need to address your question effectively, you waste time and money. At the other extreme, if your experiment requires the use of expensive consumables, or is extremely time consuming, there is a temptation to make it as small as possible. However, if your experiment is so small that it has little chance at all of allowing you to detect the effects that you are interested in, you have saved neither time nor money, and you will probably have to repeat the experiment, this time doing it properly. Problems like these can be avoided with a bit of careful thought, and in later chapters (particularly Chapter 6) we will discuss ways of doing so.

Similarly, it is not uncommon for people to collect as many different measurements on their samples as possible without really thinking about why they are doing so. At best this may mean you spend a great deal of time collecting data that you have no use for, and at worst may mean that you don't collect the information that is critical for answering your question, or that you do not give sufficient time or concentration to collecting the really important information. Don't be over-ambitious: better you get a clear answer to one question than a guess at the answers to three questions.

➡ While it is always tempting to jump into an experiment as quickly as possible, time spent planning and designing an experiment at the outset will save time and money (not to mention possible embarrassment) in the long run.

1.2.2 Ethical issues

If the only issue at stake in ill-conceived experiments was wasted effort and resources, that would be bad enough. However, life science experiments have the additional complication that they will often involve the use of animals (or animal-derived material). Experimental procedures are likely to be stressful to animals; even keeping them in the laboratory or observing them in the wild may be enough to stress them. Thus, it is our duty to make sure that our experiments are designed as carefully as possible, so that we cause the absolute minimum of stress and suffering to any animals involved. Achieving this will often mean using as few animals as possible, but again we need to be sure that our experiment is large enough to have some chance of producing a meaningful result (see Chapter 6).

There are often many different ways that a particular experiment could be carried out. A common issue is whether we apply several treatments to the same individuals, or apply different treatments to different individuals. In the former case we could probably use fewer individuals, but they would need to be kept for longer and handled more often. Is this better or worse than using more animals but keeping them for less time and handling them less often? These arguments should be weighed up before the experiment is carried out to ensure that suffering is minimized. Issues such as this will be explored in more detail in Chapter 10.

Ethical concerns do not only apply to scientists conducting experiments on animals in a laboratory. Field experiments too can have a detrimental effect on organisms in the environment that the human experimenters are intruding on. There is no reason to expect that such effects would even be confined to the organism under study. For example, a scientist sampling lichens from hilltops can disturb nesting birds or carry a pathogen from one site to another on their collection tools. For these reasons ethics will be a recurrent theme woven throughout the book.

➡ We cannot emphasize too strongly that, while wasting time and energy on badly designed experiments is foolish, causing more human or animal suffering or more disturbance to an ecosystem than is absolutely necessary is inexcusable.

1.3 The relationship between experimental design and statistics

It might come as a surprise that we are not going to talk about statistical analysis of data in any detail in this book. Does this mean that we think that statistical tests are not important to experimental design and that we don't need to think about them? Absolutely not! Ultimately, experimental design and statistics are intimately linked, and it is essential that you think about the statistics that you will use to analyse your data before you collect it. As we have already said, every statistical test will have slightly different assumptions about the sort of data that they require or the sort of hypothesis that they can test. So it is essential to be sure that the data that you collect can be analysed by a test that will examine the hypothesis that you are interested in. The only way to be sure about this is to decide in advance how you will analyse your data when you have collected it. Thus, whilst we will not dwell on statistics in detail, we will try to highlight throughout the book the points in your design where thinking about statistics is most critical.

There are two reasons why we will not concentrate on statistics in this book. The first is simply that there are already some very good statistics books available, that you can turn to when you are at the stage of deciding what test you are planning to do (see the Bibliography for a guide to some that we have used). However, the second, more important, reason is that we believe strongly that experimental design is about far more than the statistical tests that you use. This is a point that can often be lost among the details and intricacies of statistics. Designing experiments is as much about learning to think scientifically as it is about the mechanics of the statistics that we use to analyse the data once we have it. It is about having confidence in your data, and knowing that you are measuring what you think you are measuring. It is about knowing what can be concluded from a particular type of experiment and what cannot.

There are two philosophical schools of thought on statistical analysis, one emphasizing p-values associated with testing of null hypotheses, the other (Bayesian) approach emphasizing model fitting. In this book when we discuss statistics it will be in the context of null hypotheses testing, but we do not think your approach to experimental design should change in any way if you plan to take a Bayesian approach to your statistics.

 Experimental design is about the biology of the system, and that is why the best people to devise biological experiments are biologists themselves.

1.4 Why good experimental design is particularly important to life scientists

Two key concepts that crop up continually when thinking about experiments (and consequently throughout this book) are *confounding factors* and *random variation*. Indeed it might be said that the two major goals of designing experiments are to account for confounding factors and to minimize random variation. Both of these subjects will be covered in depth in Chapters 3 and 4, respectively. However, we mention them here in order to give a flavour of why experimental design is particularly important to life scientists.

1.4.1 Random variation

Random variation is also called *between-individual variation, inter-individual variation, within-treatment variation,* or *noise*. This simply quantifies the extent to which individual subjects in our sample (which could be animals, plants, humans, study plots, or tissue samples, to give but a few examples) differ from each other for reasons other than the one we are interested in. We discuss the consequences of this fully in Chapter 4. For example, all 10-year-old boys are not the same height. If our study aims to study national differences in the height of 10-year-old boys, we should expect that all boys in our study will not have the same height: and that this variation in height will be driven by

> Experimental design is about removing or controlling for variation due to factors that we are not interested in (**random variation** or noise), so we can see the effects of those factors that do interest us.

a large number of factors (e.g. diet, socioeconomic group) as well as the possible effect of the factor we are interested in (nationality). The problem for us in carrying out an investigation is that the more variation there is that is driven by other factors, the more difficult it is to detect and describe variation due to the factor we are really interested in. Thus, experimental design is often focused on reducing variation due to other factors. In our example here we probably have an interest in children more generally, but have focused only on 10-year-old boys to remove variation due to age and sex. Other sources of noise will remain, and good experimental design will help us avoid situations where that noise (variation due to other factors) drowns out the information we are interested in (variation linked to nationality).

Random variation is everywhere in biology. All the sheep in a flock are not exactly the same. Hence, if we want to describe the characteristic weight of a sheep from a particular flock, we generally cannot simply weigh one of the sheep and argue that its weight is also valid for any and all sheep in the flock. It would be better to measure the weights of a representative sample of the flock, allowing us to describe both the average weight of sheep in this flock and the extent of variation around that average. Things can be different in other branches of science: a physicist need only calculate the mass of one electron, because (unlike sheep) all electrons are the same weight. Physicists often need not concern themselves with variation, but for life scientists it is an ever-present concern, and we must take it into account when performing experiments.

→ Good experiments minimize random variation, so that any variation due to the factors of interest can be detected more easily. Statistical analysis often involves partitioning variation into parts that can be attributed to different factors, and good experimental design can make this easier.

1.4.2 Confounding factors

If we want to study the effect of variable A on variable B, but variable C also affects B, then C is a **confounding factor** (also called a *third variable*).

Often we want to understand the effect of one factor (let's call it variable A) on another factor (B). However, our ability to do this can be undermined if B is also influenced by another factor (C). In such circumstances, C is called a **confounding factor** (sometimes referred to as a *confounding variable* or *third variable*). For example, if we observe juvenile salmon in a stream to see if there is an effect of the water temperature (variable A) on rates of foraging (variable B), then the number of hours of sunlight in the day (variable C) may be a confounding factor. We might expect increased water temperature to increase foraging activity, but we might also expect increased sunlight to have the same effect. Disentangling the influences of temperature and sunlight is likely to be particularly problematic since we might expect sunlight and water temperature to be linked (it is both sunnier and warmer in summer than winter, and the amount of sunlight may actually affect water temperature), so it will be challenging to understand the effect of one in isolation from the other. Hence confounding factors pose a challenge, but not an insurmountable one, as we will discuss fully in Chapter 3.

? Q 1.3 If we are interested in comparing eyesight between smokers and non-smokers, what other factors could contribute to variation between people in the quality of their eyesight? Are any of the factors you have chosen likely to be related to someone's propensity to smoke?

Another advantage that the physicist has is that they can often deal with very simple systems. For example, they can isolate the electron that they want to measure in a vacuum chamber containing no other particles that can interact with their particle of interest. The life scientist generally studies complex systems, with many interacting factors. The weight of an individual sheep can change quite considerably over the course of the day through digestion and ingestion or through being rained on. Hence, something as simple as taking the weight of animals is not as simple as first appears. Imagine that today you measure the average weight of sheep in one flock and tomorrow you do the same for another flock. Let's further imagine that you find that the average weight is higher for the second flock. Are the animals in the second flock really intrinsically heavier? Because of the way that you designed your experiment it is hard to know. The problem is that you have introduced a confounding factor: time of measurement. The sheep in the two flocks differ in an unintended way, in that they were measured on different days. If it rained overnight then there might be no real difference between sheep in the two flocks, it might just be that all sheep (in both flocks) are heavier on the second day because they are wetter. We need to find ways to perform experiments that avoid or account for confounding factors, so that we can understand the effects of the factors that we are interested in. Techniques for doing this will be tackled in Chapters 7–10. However, before we can do this, we have to be clear about the scientific question that we want a given experiment to address. Designing the right experiment to answer a specific question is what the next chapter is all about.

Q 1.4 Faced with two flocks of sheep 25 km apart, how might you go about measuring sample masses in such a way as to reduce or remove the effect of time of measurement as a confounding factor?

 Confounding factors make it difficult for us to interpret our results, but their effects can be eliminated or controlled by good design.

1.4.3 Simpson's paradox

We need to be alert to confounding factors not just when designing experiments, but also when interpreting any data set. Named after the British statistician Edward H. Simpson, Simpson's paradox is an example of the dangers of uncritically pooling data without considering confounding factors. To illustrate the problem we will consider the case of a US university that was accused of having an admissions policy biased against women. The example is based on a real case, although the numbers presented are made up to make the example easier to follow. Imagine we examined a university's admissions figures for one year and found that of the 8441 men that applied 44% were accepted, whilst of the 4321 women that applied only 35% were accepted. It is clear from the data that a man who applied to the university was more likely to be accepted than a woman.

However, we should pause before interpreting this as evidence of sexist selection procedures. Universities offer many kinds of courses and we might question the wisdom of pooling the data across all of these courses. So let's look at the same data in another way, this time breaking it down into two broad subject areas. These data are given in Table 1.1.

Gender of applicant	Outcome of application	
	Success	**Fail**
Subject area A		
Male	2954 (50%)	2955 (50%)
Female	337 (52%)	311 (48%)
Subject area B		
Male	760 (30%)	1772 (70%)
Female	1175 (32%)	2498 (68%)

Table 1.1 Breaking down the data

When we look at the data broken down by subject area, then there seems little difference in the percentages of men and women accepted within each subject area; indeed, if anything, for both subject areas women are slightly more successful than men.

This does seem paradoxical. The conclusion we draw from the data pooled across the university is the opposite of that drawn if we consider the data for the two subject classes separately. The explanation for the paradox is that women and men differ in the subjects that they apply to study, whilst the subject areas vary considerably in their acceptance rates. Only 15% of women apply for subject area A, which has a high acceptance rates (for both sexes), whilst 70% of men apply for subjects in this area. Whilst there is no doubt that it is harder for women to get into this university than men, the reason appears to be that women choose courses that are harder to get into,

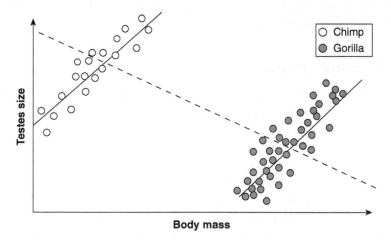

Figure 1.1 An example of Simpson's paradox with continuous data: body mass on the x-axis and testes size on the y-axis. If you ignore the issue of the species of individuals then it appears that testes size decreases with body size. Actually, testes size increases with body mass in both chimpanzees and gorillas, but because the degree of sexual competition amongst males differs between the two species, they invest differently in sperm production. Chimpanzees are smaller in body mass but have greater testes size than gorillas.

rather than an inherent sexism in selection procedures. Simpson's paradox illustrates clearly the potential hazards of uncritically pooling data without critically considering confounding factors which might be important.

Whilst our example is based on discrete data, Simpson's paradox can also occur in continuous data and an example of this is shown in Figure 1.1.

➜ Think very carefully about potential confounding factors before interpreting data. Failure to do so can lead to highly inappropriate conclusions about causal factors.

1.5 **Subjects, experimental units, samples, and terminology**

Experimental design, like any other area of study, has built up its own terminology, and our aim throughout this book (and sometimes in our supplementary material) is to try and demystify any terminology that you might come across when reading up on experimental design. We want to finish this chapter by focusing on an area of experimental design where a bewildering range of specialist terms are used for what at first seems a very straightforward concept. In general, a scientific study will involve you taking measurements on a sample of 'things': you might be measuring lengths of strands of DNA, diameters of cells, stress hormone levels in samples of blood, feeding rates in fish, number of breeding pairs in penguin colonies, or fraction of the human population that is under 14 years old in entire countries. So the 'things' that make up our sample can be very diverse. In this book, where we need a name for the 'things' in our sample, we will generally call them *subjects*, but you might also encounter them called *experimental units*, *individuals*, *replicates*, or *participants*.

A complication here is that sometimes what exactly we regard as the subject can change in different parts of a study, and this can add more terminology. Sometimes practical or ethical considerations argue for keeping subjects together for at least part of a study. For example, many rodents are highly social; keeping such an animal alone in a cage would cause it stress. Keeping it alone might also cause it to behave quite differently from how it would naturally in a group in ways that make the experiment difficult to generalize to animals under more natural conditions. Practically, you might often have four fish tanks, each of which could ethically hold twenty fish of your study species; but you are rarely going to have the luxury of eighty tanks to hold all the fish in separately. In a study of how different educational techniques work on children, we have to bear in mind that children generally receive their education in a group setting in a school classroom.

Often in an experiment we will want to randomly allocate subjects to different experimental treatments. There are two possibilities: each subject could be randomized such that a group contains subjects allocated to different experimental treatments, or randomization could occur at group level so that all of the members of a group are allocated to the same treatment. Sometimes practical aspects will dictate that allocation must be done at a group level. For example, if we are investigating how cage size affects mouse behaviour then clearly the cage size has to be the same for all of the mice in the cage. When comparing two different ways to teach children a foreign language it might be practically much more convenient to allocate an entire class to a given teaching style, rather than split classes randomly and have the two half-classes taught separately for that subject. However, for other manipulations mixed groups seem possible, for example if the manipulation is either a hormone injection or a placebo injection for the mice; or if children are allocated to wearing either school uniform or their own choice of clothes. In these cases, the experimenter has the choice of randomizing at the individual level or the group level. In the jargon of experimental design: the *unit of randomization* could be either the mouse or the cage of mice; or could be either the child or the class of children.

Similar issues arise with the level at which we take measurements. The *unit of observation* (sometimes called the *unit of evaluation* or *unit of sampling*) could be either the individual or the group. Generally the unit of observation will be obvious to you from the nature of the measurement, and will normally but not always be the individual. If we are interested in how the hormone affects sleeping patterns then our level of measurement could be the individual (providing we have mice individually identified). We simply inspect the cage at regular intervals and note which individuals are asleep. Similarly, for language teaching methods, our level of observation might naturally be the individual; since we measure learning most obviously by asking each child to complete a standardized test paper. However, sometimes the unit of observation is naturally the group. For the mice we might be interested in how the hormone affects drinking behaviour and the most natural way to do this is to note change in the level of the water in the reservoir of the single drinker provided in each cage for communal use. But in this case a finer resolution at the level of individual mice would be possible if the mice were made individually identifiable visually and a video camera was trained on the drinker. Our advice here is that you should adopt the most fine-grained unit of observation that is practically possible. In the case of the effect of hormones on mouse water consumption, you might decide that the hours of watching video tape are not a good use of your time. You might then have the cage as the *unit of observation* in your study even though individual-scale observation was technically possible, on grounds of rational use of your time.

Now we introduce the final piece of jargon: *the unit of analysis*. This is the level at which we carry out our statistical analysis of the data. Generally this will be the same as the unit of observation, but we might sometimes aggregate to a more broad scale (measuring each mouse separately but working with the average value for each cage rather than each mouse) in our statistics in order to be more confident that the units of analysis meets the statistical requirement that the values of each subject are independent (something we will discuss fully in Chapter 5). Clearly, the analysis can never be more fine-grained than the unit of measurement: that is we can never use measurements taken at the level of a group to make inferences about individual members of the group (doing so is called the *Ecological Fallacy*).

The *Ecological Fallacy* involves carrying out an analysis at a group level but misinterpreting the results as if they applied at an individual level. A famous example of this is a study that found a negative relationship across US states between the fraction of a state's population that were immigrants and the fraction of the state's population that were illiterate. That is, states with lower illiteracy levels tended to have a higher fraction of the population who were immigrants. This was interpreted as evidence that immigrants were less likely to be illiterate than those born in the USA. However this interpretation was wrong. In fact, in each state, immigrants were more likely to be illiterate, but immigrants were also more likely to settle in some states than others and particularly more likely to settle in states that happened to have low levels of illiteracy. In addition, although the fraction of the population that were immigrants varied from state to state, these fractions where generally low; and so the immigrant population had little effect on a particular state's level of illiteracy. The mistake was to use data taken at the state level, but draw conclusions about individuals.

In order to give us a simple term for the 'things' in a sample, we will generally use the word 'subject', to mean either unit of randomization, unit of observation, or unit of analysis. Which of these we are referring to should always be clear from the context, and in particular which part of the process of a scientific study we discussing: either setting up the experiment, collecting measurements, or analysing the data.

If you feel the need to cement any of the issues raised in this chapter, then you can find some further self-test questions in the supplementary information; otherwise let's push on. In the next chapter we explore how being clear about the scientific question that we want a given study to address will help us design the study effectively.

➡ In most scientific studies what the subjects in your sample are will be obvious and uncontroversial, but in some studies these will change for different stages of the study.

◼ Summary

- You cannot be a good life scientist without understanding the basics of experimental design.

- The basics of experimental design amount to a small number of simple rules; you do not have to get involved in complex mathematics in order to design simple but effective experiments.

- If you design poor experiments, then you will pay in time and resources wasted.

- Time and resource concerns are trivial compared to the imperative of designing good experiments so as to reduce (or hopefully eliminate) costs to your experiment in terms of suffering to animals or humans or disturbance to an ecosystem.

- Confounding factors make it difficult for us to interpret our results, but their effects can be eliminated or controlled by good design.
- Think very carefully about potential confounding factors before interpreting data. Failure to do so can lead to highly inappropriate conclusions about causal factors.
- In most scientific studies what the subjects in your sample are will be obvious and uncontroversial, but in some studies these will change for different stages of the study.

Starting with a well-defined hypothesis

2

Our objective in this chapter is to help you see the need to carefully define the aims of your study in order to come up with a good design to meet these aims.

- Your aim in conducting a scientific study is to test one or more specific scientific hypotheses.
- There is no way that you can hope to design a good study unless you first have a well-defined hypothesis (see Section 2.1).
- One aim of good design is to produce the strongest test of the hypothesis (Section 2.2).
- Careful selection of a control group can be essential to designing an experiment to address exactly the question you are interested in (Section 2.3).
- A pilot study will allow you to focus your aims and perfect your data-collection techniques (Section 2.4).

2.1 Why your study should be focused: questions, hypotheses, and predictions

A **hypothesis** is a clear statement articulating a plausible candidate explanation for observations. It should be constructed in such a way as to allow gathering of data that can be used to either refute or support this candidate explanation.

There can be several hypotheses for the same observation. For example, take the observation that patients in the beds nearest to the entrance to a hospital ward respond more effectively to treatment than those at the other end of the ward. Candidate hypotheses could include:

1) Patients are not distributed randomly to beds, with the ward sister having a tendency to place more severe cases at the quieter end of the room away from the entrance.

> A **hypothesis** is a clearly stated postulated description of how an experimental system works.

Q 2.1 Suggest some hypotheses that could explain the observation that people drive faster on the journey to work than on the way home.

2) Patients nearer the door are seen first on the doctor's round and end up getting more of the doctor's time and attention.

3) Patients respond positively to increased levels of activity and human contact associated with being nearer the ward entrance.

Testing specific hypotheses helps give focus to your study and increases your chances of actually finding out something interesting. Don't get yourself into a position where you think something like:

'Chimps are really interesting animals, so I'll go down to the zoo, video the chimps, and there are bound to be lots of interesting data in 100 hours of footage.'

A **pilot study** is an exploration of the study system conducted before the main body of data collection in order to refine research aims and data-collection techniques. A good pilot study will maximize the benefits of your main data-collection phase and help you avoid pitfalls.

We agree that chimpanzees can be very interesting to watch, so by all means watch them for a while. However, use what you see in this exploratory **pilot study** to generate clear questions. Once you have your question in mind, you should try to form hypotheses that might answer the question. You then need to make predictions about things you would expect to observe if your hypothesis were true and/or things that you would not expect to see if the hypothesis were true. You then need to decide what data you need to collect to either confirm or refute your predictions. Only then can you design your study and collect your data. This approach of having specific questions, hypotheses, and predictions is much more likely to produce firm conclusions than data collected with no specific plan for its use. This should be no surprise; if you don't know how you are going to use data, the likelihood of you collecting a suitable amount of the right sort of data is low.

See Section 2.1 of the supplementary material for a more in-depth comparison between exploratory and confirmatory studies. Go to: **www.oxfordtextbooks.co.uk/ orc/ruxton4e/**.

Given the importance of a focused research question, you might be thinking that spending even a few minutes making unfocused observations of the chimps is simply wasted time that could be better spent collecting data to answer scientific questions. We disagree. You should be striving to address the most interesting scientific questions you can, not as many questions as you can. Darwin, Einstein, and Watson and Crick are not famous because they solved more problems than anyone else, but because they solved important problems. So take the time to think about what the interesting questions are. We are not promising that you'll become as famous as Einstein if you follow this advice, but the chances will increase.

As a general rule your experiment should be designed to test at least one clear hypothesis about your research system. A period of unfocused observation can help you identify interesting research questions, but a study based entirely on unfocused observation will rarely answer such questions definitively.

2.1.1 An example of moving from a question to hypotheses, and then to an experimental design

The last section might have seemed a little abstract, so let us try an example. Imagine that you have collected a number of carcasses of black grouse (a game bird) as part of a study looking at their mortality risk due to flying into man-made structures such as ski-lifts, wind turbines, and high-voltage electricity cables. You notice that males seem to have higher ectoparasite loads than females. This leads naturally to the research question:

Why do males have higher parasite loads than females?

Our observations have allowed us to come up with a clear question, and with a clear question we have at least a fighting chance of coming up with a clear answer. The next step is to come up with hypotheses that could potentially explain our observation. One such hypothesis might be:

The stress associated with competition for mates reduces males' ability to mount defences against parasites.

Now how would we test this hypothesis? The key is to come up with a number of predictions of observations that we would expect to make if our hypothesis is correct. So a prediction has to follow logically from the hypothesis *and* has to be something we can test. In the case of this hypothesis, we might make the prediction:

Parasite loads are positively correlated with levels of circulating stress hormones, and males show higher levels of these hormones than females.

By making a clear prediction like this, it becomes obvious what data we need to collect to test our prediction. Data that support our prediction provide some support for the hypothesis (assuming that we have come up with a sensible prediction that tests our hypothesis). In statistics texts, you'll find hypotheses linked as pairs with each pair being composed of a *null hypothesis* and an *alternative hypothesis*. The null hypothesis can be thought of as the hypothesis that nothing is going on. In the example here, a possible null hypothesis would be:

Parasite loads are unrelated to stress.

Such a null hypothesis would lead to the prediction:

There is no relationship across individuals between parasite load and level of circulating stress hormones.

The hypothesis that we originally gave is an example of an alternative hypothesis, i.e. a hypothesis in which the pattern that we see is due to something biologically interesting, rather than just being due to chance. The reason that we need a null hypothesis is right at the heart of the way in which statistics works, but the details needn't bother us too much here. However, it boils down to the fact that most statistical tests work by testing the null hypothesis. If our observations allow us to reject the null hypothesis (because the data we collect disagrees with a prediction based on the null hypothesis), then this gives us support for our alternative hypothesis. So if we can reject the null hypothesis that chimp activity is not affected by feeding regime, logically we should accept that chimp activity is affected by feeding regime.

This may seem rather a twisted way to approach problems but there is good reason behind this approach. Philosophically, science works conservatively. We assume that nothing interesting is happening unless we have definite evidence to the contrary. This stops us getting carried away with wild speculations, and building castles in the air. This philosophy leads to concentration on the null hypothesis rather than the alternative hypothesis. However, whilst this philosophy is the bedrock of statistical testing, it need not detain us further here. Just remember that for every hypothesis you make that suggests that something interesting is happening there will be a corresponding null hypothesis (that nothing interesting is happening), and that by testing that null hypothesis you can gain information about the hypothesis that you are interested in.

 In order for your hypothesis to be useful it has to generate *testable* predictions.

2.1.2 An example of multiple hypotheses

There is nothing to say that you are limited to a single hypothesis in any particular study. Let's consider a second example: you are walking along a rocky shore and you notice that the whelks (sea snails) on the rocks often seem to occur in groups. You come up with the sensible question:

   Why do whelks group?

Your first thought is that it might be to do with seeking shelter from the mechanical stresses imposed by breaking waves, so you generate a hypothesis:

   Whelks group for protection from wave action.

Then, after a little more thought, you wonder if maybe grouping is something to do with clumping in areas of high food, and this leads you to a second hypothesis:

   Whelks group though feeding.

As with the previous example, we now need to come up with predictions that we can test. Ideally, we would like predictions that will allow us to discriminate between these two hypotheses. If the first hypothesis is true, we might predict:

   Whelks are less likely to be found in groups in areas with low wave action.

Whereas if the second is true our prediction would be:

   Whelks are more likely to be found in groups in areas of higher food density.

With these predictions in hand we could then design a study to test both of these predictions. Of course, even if we find one of our predictions to be supported, for example, that whelks are less likely to be found in groups in areas with low wave action, this does not necessarily mean that the hypothesis that led to the prediction ('whelks group for protection from wave action') is correct. Someone might come up with another hypothesis that would also lead to this prediction. For example, the hypothesis:

   Whelks are more vulnerable to predators in areas with strong wave action, but grouping provides protection from predators.

could also lead to the prediction:

Whelks are less likely to be found in groups in areas sheltered from wave action.

Your task will then be to come up with another prediction that can discriminate the new hypothesis from the original one. Now that is not to say that both hypotheses cannot be true. Maybe part of the reason that we find aggregations of whelks in wave-exposed areas is to mitigate the mechanical effects of wave action, and part of it is to gain protection from predators that are more active in wave-exposed areas. Indeed, it is common in life sciences to find that several hypotheses need to be combined to fully explain observations. The key thing is that if only one hypothesis is correct, we should be able to determine that from our study. To clarify what we mean let's call these two hypotheses *predation* and *shelter*. We then have four possibilities for how the system actually works.

Possibility 1: Neither hypothesis is true and the observed patterns are due to something else entirely.

Possibility 2: *Predation* is true and *shelter* is false.

Possibility 3: *Shelter* is true and *predation* is false.

Possibility 4: Both *predation* and *shelter* are true.

A study that allowed us to discriminate possibility 1 from 2, 3, and 4, but did not allow discrimination between 2, 3, and 4 would be useful, but not ideal. The ideal study would allow us not only to discriminate possibility 1 from the others, but also discriminate between these other possibilities. That is, at the end of the study, we could identify which of the four possible combinations of the two hypotheses is actually correct. Such a study might include a series of experimental treatments like those shown in Figure 2.1, where either predators or wave action or both were removed and the grouping behaviour of the whelks observed. Alternatively, it might involve making observations on beaches with different combinations of predation and wave action. No matter how the study is organized, the important thing is that the best study will be the one that allows us to tease apart the influence of the different hypothesized influences on grouping behaviour.

Generating sensible predictions is one of the 'arts' of experimental design. Good predictions will follow logically from the hypothesis we wish to test, and not from other rival hypotheses. Good predictions will also lead to obvious studies that allow the prediction to be tested. Going through the procedure of:

question → hypothesis → prediction

is an extremely useful thing to do, in that it makes the logic behind our study explicit. It allows us to think very clearly about what data we need to test our predictions, and so evaluate our hypothesis. It also makes the logic behind the study very clear to other people. We would recommend strongly that you get into the habit of thinking about your scientific studies this way.

Biological systems are complicated, so it is common in life sciences to find that several hypotheses need to be combined to fully explain a set of observations.

Q 2.2 A report claims that a questionnaire survey demonstrates that those who reported regularly playing computer games displayed attitudes consistent with an increased propensity to perform violent acts. Suggest some hypotheses that could explain this association and how predictions from these hypotheses might be tested.

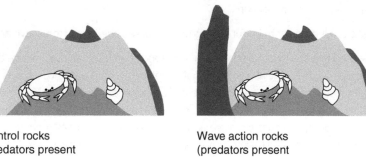

Control rocks
(predators present
normal wave action)

Wave action rocks
(predators present
wave action reduced)

Predator effect rocks
(predators excluded
normal wave action)

Combined rocks
(predators excluded
wave action reduced by
barrier on the left)

Figure 2.1 Experimentally treated rocks to examine whether predation, wave action, or both affect the grouping behaviour of whelks. The real study would use multiple rocks of each type for reasons outlined fully in Chapter 4.

2.1.3 Where do (good) ideas come from in the first place?

This is tricky, and there is no golden secret—or if there is, we don't know what it is—but here is some advice that might help.

Read a lot and listen to scientists presenting their work; but read and listen critically. As you read or listen keep in mind the following questions:

- Can you see any potential flaws in the work?
- What are the limitations to the study?
- How could those limitations be overcome?
- What would we learn from overcoming those limitations over and above what is already known from this study?
- How widely applicable are the results, to other species, or contexts?
- What is the logical next step in this line of research?
- Does this work have applicability to other questions that the authors have not realized?

Keep a notebook or computer file in which you record your thoughts. Not only does this mean you can come back to these ideas in the future, but the act of committing the ideas to paper will help you to focus and clarify them further. Learning to think critically in this way should provide a fund of ideas. However, these ideas may often be solid but unspectacular developments of previous work; what if you want to come up with really novel ideas?

Perhaps the key to having really novel ideas is just to keep your eyes and ears open and try and question the things you see around you. Maybe you'll see some surprising behaviour from squirrels next time you are walking through the park, and some internet searching that evening does not explain the behaviour satisfactorily to you. When a colleague with back problems reports to you that they have recently become a father for the first time, you ask yourself whether such an event might be related to unusual lifting behaviours or sleeping patterns that might underlie the back problem. Richard Feynman, one of the greatest physicists of the last century, told how wondering why a plate that had been thrown across a canteen wobbled in a particular way as it flew led him to important insights (and in part to his Nobel Prize in Physics). Ideas can come from the most unlikely places. Some people have a naturally curious mind and a gift for creative thinking; the rest of us can improve our performance with practice. Practice is a lot more fun and effective with other people involved, so talk to other people about things you are currently puzzling over. They don't have to be experts, just people whose intellect you respect. And when you see someone present their work, don't be afraid to ask questions. It is easy to assume that your question is not interesting, or that you have simply misunderstood something and you will appear foolish in front of your peers. If you really have misunderstood something, then you are probably not alone and the speaker will value the opportunity to clarify. However, perhaps your question is really significant, uncovering a problem that the speaker had not been aware of, or links to other areas of research which will benefit their work. If you don't ask you will never know (or worse, the senior professor in your department will ask the same question, impressing everyone with their brilliance and insight!)

These are ways to generate ideas, and you can generate a lot of ideas, more than you could ever follow up; so the next challenge is to decide which are the best ideas. First of all, ask yourself whether it is actually possible for you to explore this question and obtain a valuable answer. If the answer is 'no', you can put that idea away in a pending file (in case circumstances change in the future), and move on to other questions. If it seems feasible to tackle a question, then you need to ask yourself whether you personally are really excited about tackling it. If you are not really excited at the start of a project, then you are probably better to move on to another one. If you are really excited about a few ideas then pitch them all to some people that you respect and see what reaction you get. If there is a project that consistently excites people and which you get more and more enthusiastic about every time you pitch it, then it's a good idea—go for it! Presenting to others should not only help you decide on the best idea, but should help you hone your thinking on that idea.

 Read and listen critically; and don't keep your ideas to yourself, bounce them off people you respect.

2.2 Producing the strongest evidence with which to challenge a hypothesis

Try to find the strongest test of a scientific hypothesis that you can. Consider the hypothesis:

> Students enjoy the course in human anatomy more than the course in experimental design.

You might argue that one way to test this hypothesis would be to look at the exam results in the two courses: if the students get higher grades in one particular course compared to another, then they probably enjoyed it more, and so were more motivated to study for the exam. Thus our prediction, based on our hypothesis, is:

> Students will get higher marks in the human anatomy exam than in the experimental design exam.

We would describe this as a weak test of the hypothesis. Imagine that students do score significantly higher in human anatomy. This is *consistent* with the hypothesis. However, it could also be explained by lots of other effects: it may be that the human anatomy exam was just easier, or that it fell at the start of the set of exams when students were fresh, whereas experimental design came at the end when they were exhausted. This problem of alternative explanations is explored further in Section 2.2.3.

 Data consistent with a hypothesis are less useful if those data are also consistent with other plausible hypotheses.

2.2.1 Indirect measures

An **indirect measure** is a measure taken on a variable that we are not primarily interested in but which can be used as an indicator of the state of another variable that is difficult or impossible to measure.

The problem with using exam scores to infer something about students' enjoyment levels is that you are using an **indirect measure** (sometimes called a *surrogate variable*, *surrogate measure*, or *surrogate marker*). This involves measuring one variable as a surrogate for another (generally more difficult-to-measure) variable. For example, if we are interested in comparing the amount of milk provided to seal pups by mothers of different ages, then we might consider that the amount of time that pups spend suckling is a reasonable surrogate for the quantity of milk transferred. Although this assumption makes data collection much easier, we must be sure that we are using a good surrogate. If rate of milk delivery during suckling varies between mothers, then time spent suckling could be a poor measure of actual quantity of milk transferred. For this reason, unless we have proof, we cannot expect others to be convinced by our assumption that milk delivery rate is a constant. You could imagine that age of the mother, her health, time since the last feed, and duration of the current feed could all plausibly influence instantaneous delivery rate.

Indirect measures generally produce unclear results, as in the case of students' exam performance discussed earlier. If we can, we should always try and measure the thing we are interested in directly. In the course-enjoyment study, we should interview the students (separately, see Chapter 5 on independence of data points) and ask them which course they prefer. However, if the questioner were the human anatomy lecturer, then we should worry about how honest the answers would be. It would be better to have the questions asked by someone who the students perceive to be disinterested in the answer.

 Don't use indirect measures unless you have to, and interpret them with care when you must.

2.2.2 Considering all possible outcomes of an experiment

One moral to be taken from this section so far is that before doing an experiment you should always ask yourself, for *every possible outcome*, how such a set of results could be interpreted in terms of the hypotheses being tested. Try to avoid experiments that can produce potential outcomes that you cannot interpret. It doesn't make you sound very impressive as a scientist if you have to report that:

> I collected data this way, the data look like this, and I have no idea what I should conclude about my study system from this.

It can sometimes be difficult to think of every possible outcome of an experiment beforehand, and again, pilot studies can be an effective way of uncovering these.

Another aspect of this warning is to be wary of doing experiments where a useful outcome of the work hangs on getting a specific result. There is nothing wrong with thinking, 'Let's do this experiment, because if our hypothesis is supported by the experiment, then we'll really cause a sensation'. However, make sure that your results are still interesting and useful if the outcome is not the one that you hoped for.

Imagine that your experiment is looking to explore the effect of one factor on another. What if you find no effect at all? You must ask yourself whether someone else would still judge your experiment to have been scientifically valid and a useful way to spend time and resources even if you get this negative result. For example, imagine that you performed an effective investigation to test the hypothesis:

> Cannabis use has an effect on driving ability.

If you find no relationship between the two, then that would be an interesting result. It is interesting because it conflicts with our current understanding of the effects of this substance. It would be relevant to both our fundamental understanding of the effect of the drug on the brain and to the scientific underpinning of motoring law related to drug use. If instead you find an effect, and can quantify the effect, then this should be an interesting result too, again with implications for our understanding of brain function and for legislation.

Now imagine that you performed an effective investigation to test the hypothesis:

> Preference for butter over margarine is linked to driving ability.

If you find strong evidence supporting this hypothesis, then this would be surprising and interesting, as it conflicts with our current understanding of the factors affecting

 See Section 2.2.1 of the supplementary material for further discussion of the special design challenges associate with collecting data from humans, including consideration of the use of deception and of randomized response techniques. Go to: **www.oxfordtextbooks.co.uk/orc/ruxton4e/**.

Q 2.3 In the original study referred to in Q 2.2, the investigators were interested in people's propensity to commit violent acts. However they used an indirect measure of this by asking people to answer a carefully designed questionnaire about their attitude to certain situations such that these answers could be used to infer propensity to violence. Why do you think this approach was taken rather than using a direct measure of propensity to violence?

driving ability. However, if you find no relationship between whether people prefer butter or margarine and their driving ability, then this is a less interesting result. Others could say that you were never going to find an effect between two such entirely unrelated factors and that you should have spent your time on more productive lines of enquiry.

We are not saying, 'don't ever do an experiment where one potential result of the experiment is uninteresting'. If you had demonstrated that a preference for butter is associated with better driving, then that could be very interesting indeed. You might consider that the fame and fortune associated with discovering such results are worth the risk of generating lots of dull results. It's your decision; our job was to point out to you that there are risks.

A common mistake is to feel that only experiments where the results show very strong relationships are interesting. This is not only flawed thinking but dangerous thinking. Desire to achieve a specific result from your experiment may lead you to be a less reliable recorder of information (see Section 11.6.3). We encourage you to design experiments that are interesting because of the question they ask more than because of the specific answer to the question that emerges from the data.

 Think about how you would interpret every possible outcome of your experiment, before you decide to do the experiment.

2.2.3 Satisfying sceptics

One theme running through this book is that your experimental design and data collection should satisfy someone that we describe as 'the Devil's Advocate'. By this, we mean that you must make sure that your experiment is designed so that the conclusions that you draw are as strong as possible. Let's imagine that you are in a situation where a group of observations could be explained by either of two contradictory mechanisms (A or B), and your aim is to try and illuminate which is in fact the real mechanism underlying the results. Don't be satisfied with an experiment that will let you conclude:

> The results of this experiment strongly suggest that mechanism A is operating. Whilst it is true that the results are equally as compatible with a mechanism B, we feel that A is much more likely.

You have to expect the person refereeing your manuscript or marking your project report to be a 'Devil's Advocate', who constantly says, 'No, I'm not prepared to give you the benefit of the doubt, prove it'. For the case described, their conclusion is:

> It is a matter of opinion whether mechanism A or B is more likely, hence I can conclude nothing from your experiment.

It would be much better to design an experiment that would allow you to conclude something stronger, such as:

> These results are entirely consistent with the hypothesis that mechanism A is in operation. However, they are totally inconsistent with mechanism B. Therefore we conclude that mechanism A is the one acting.

The Devil's Advocate is wilfully looking to shoot holes in your argument—you must restrict their ammunition. The Devil's Advocate cannot argue that mechanism B is operating if you have proved that it is not.

➡️ **You should think of the Devil's Advocate as a highly intelligent but sceptical person. If there is a weakness in your argument, then they will find it. However, they can be made to believe you, but only when they have no reasonable alternative. They will not give you the benefit of any reasonable doubt.**

2.3 Controls

Often your research question involves comparing individual experimental units belonging to different groups. Often one of the groups is not especially interesting in itself, but is a **control group** that offers a baseline against which the values of individuals in other groups can be compared. Careful selection of this control group is essential to making sure that your experiment answers exactly the question you want it to answer. Suppose that we claim that buying this book will improve your performance in subsequent exams. You decide that anything that will improve your marks has to be worth such a modest investment, and so buy the book (and maybe even read it). Subsequently, when the exam results are announced, you are delighted to find that you have done better than you expected. Does this mean that our claim was true? Well, you may feel that the book helped you to design better experiments or improved your ability to critically read the literature (or helped you get to sleep at night). Indeed, we hope there are many reasons why this book will help to improve your scientific abilities. However, based only on your experience it is hard to say that the book had any effect on your exam performance at all. The reason is that we don't know what scores you would have got if you hadn't bought the book. Your experiment to examine the effect of buying this book on your exam marks is flawed because it has no **control**.

2.3.1 Different types of control

Most, but not all (see Section 2.3.4), experiments require a control. If you have followed our advice by clearly stating the specific hypothesis under test, then identifying whether a control is necessary, and what that control should be, is relatively easy. Consider the following hypothesis:

> Giving a vitamin supplement with the food of caged rats leads to increased longevity.

In order to test this, we need to measure the longevity of rats to which we give the vitamin supplement. However, this alone would not be enough to answer the question, because we don't know how long the rats would have lived if they had not been given the supplement. We need a control. In an ideal world, we would compare the longevity of the same rats both when they have been given a supplement, and when they have

A **control** is a reference against which the results of an experimental manipulation can be compared. For example, if we want to explore the effect of smoking on human lung tissue, we need a **control group** of tissue samples from non-smokers to compare with our **treatment group** of tissue samples from smokers.

not. However, a rat can only die once, so clearly this is not an option. Instead we need a *control group* of rats that are identical in every way to our **treatment group** (sometimes called *experimental group*), except that they do not experience the experimental manipulation itself. By 'identical' we mean in every practical way: the rats must be of the same strain, from the same source, kept under the same conditions. This is easily done by randomly assigning individuals from a common pool to the two groups: *treatment* and *control* (see Chapter 7). The control group allows us to estimate the effect of the supplement, because we can measure what happens without the supplement.

Careful choice of exactly the correct control is vital to good experimental design. To help you appreciate this, we need to introduce some jargon: **positive and negative controls**.

The rat-longevity study just discussed used what we would class as a negative control, because we expect the manipulation experienced by control individuals to have no effect on the factor of interest (longevity). A commonly encountered example of a negative control occurs in drug trials where subjects in the control group are given pills that are otherwise identical to those used to deliver the active ingredient under investigation but entirely lack that ingredient. These are often called a *placebo* pill or *vehicle control*.

A second kind of control that is needed in some studies is called a positive control. This is a treatment group where we expect to see an effect on the thing that we are measuring. A common use of positive controls is to confirm that our experimental procedures are working. So imagine that you are testing bacterial isolates collected in your local hospital for resistance to methicillin. You spread a sample of each isolate in an agar plate containing the antibiotic, and then see whether the bacteria can grow. When you return to the lab the following day you are pleased to find that not one of the isolates has been able to grow in the presence of the methicillin. However, before you give the hospital a clean bill of health think about what these results might mean. One explanation is that the new infection control procedures introduced by the hospital management have worked and resistance is essentially absent. However, there is another possibility ... perhaps your experimental protocol has not worked correctly. Maybe you forgot to add some critical component to your growth media that means it cannot support bacterial growth, or maybe there was a problem with your incubator overnight that meant the temperature dropped for a long period of time slowing the growth to levels that you cannot measure. These possibilities could have been ruled out if you had included samples of bacteria that are known to be resistant to methicillin in your study as a positive control. If, as expected, these bacteria are able to grow then you can be confident that you have set things up correctly, and your results are meaningful. However, if these bacteria are also unable to grow then you know you cannot interpret the results of the rest of the study. As an aside, you would also want a negative control of sensitive bacteria to ensure that your methicillin is working correctly (or that you did not forget to add it!)

Positive controls can be used in other ways too. Consider the hypothesis:

> A particular novel treatment for warts in humans produces better results than the currently used method.

Very often, a control group is a group to which a manipulation is applied that is expected to have no effect. This is called a **negative control**. However, sometimes, the research question will call for the control group to be manipulated in some way that we expect to cause an effect (generally one that is known and understood); this is called a **positive control**.

In this case, we must measure the clear-up rates of warts on a group of patients given the novel treatment. However, to compare these rates with those of patients given no treatment at all would not test the hypothesis posed; rather it would address the hypothesis:

The novel method produces a beneficial effect on the clear-up of warts, compared to taking no action.

To address the original hypothesis, we need a control group of patients that are given the established method of treatment, rather than no treatment at all.

Of course, as with the methicillin study, there is no reason why this experiment cannot have both types of control. If we just have the positive control of treating people with the new treatment or the established treatment, we can say which treatment is better. It is unlikely, but possible, that both treatments actually make warts worse, and one is just less bad than the other. The inclusion of a third, negative control, group would allow this possibility to be examined, and (if the results turned out that way) for us to say with confidence that the new method was not only less bad at removing warts than the existing treatment, but that both were actually better than not doing anything at all. Further, we could quantify how much better they performed relative to simply taking no action at all.

Is a control really necessary in this example? You might argue that there are probably already studies on the effectiveness of the existing treatment. Could we not forget about using our own control and simply compare the results of our treatment group of patients with the published results of previous studies of the effectiveness of the established method? In the jargon, this would be using a **historical control** rather than a **concurrent control**. This approach has the attractions of requiring less effort and expense, two worthwhile aims in any study. However, there are potentially very serious drawbacks. The key thing about the control group must be that it differs from the treatment group in no way except for the treatment being tested (this is to avoid confounding factors). This is easier to achieve with a concurrent control than a historical one. Imagine that we found the new treatment cleared people's warts in an average of two weeks, while a study carried out three years ago in another hospital found that the original treatment cleared warts in four weeks. It might be that the difference between these two studies is due to the differences in treatment. However, it could also be due to anything else that differed systematically between the studies. Maybe the individuals in one study had more serious warts, or a different type of wart. What if the techniques of applying the treatments differed, or the definitions of when warts had cleared up differed? All of these other confounding variables mean that we can have little confidence that the apparent differences between the two trials are really due to the different treatments. We may have saved effort and expense by using a historical control, but with little confidence in our results, this is clearly a false economy. A historical control can certainly save time and resources, and may have ethical attractions, but have a long think about how valid a control it is for your study before deciding to adopt a historical rather than a concurrent control.

If we run our own control group at the same time as the experimentally manipulated group, this is called a (positive or negative) **concurrent control**; if instead we use historical data as a reference, this is called a **historical control**.

Q 2.4 One of us once heard a colleague say that they were pleased to have finished their experimental work, and that all they had left to do was to run their control groups. Should they be worried?

 Careful statement of the hypothesis under test makes it easy to determine what type of control your experiment requires.

2.3.2 Making sure the control is as effective as possible

If we want to have the maximum confidence in our experimental results, then the control manipulation should be as similar to the experimental manipulation as possible. Sometimes this will require some careful thought into exactly what control we need. Otherwise we might not avoid all possible confounding factors.

For example, consider an experiment to see whether incubating an artificially enlarged clutch of eggs changes the nest attentiveness of a parent bird. We will need a group of control nests in which the clutch size is not enlarged. Let us imagine that these controls are simply nests that are monitored in the same way as the treatment group of nests but which were not subjected to any manipulation from the experimenters. In contrast, the experimenter added another egg to each of the treatment nests before the observation period. If we now find a difference in parental behaviour in the treatment group compared to the controls, then we cannot safely attribute this to the difference in clutch size. The Devil's Advocate might argue that the difference found between the two groups is nothing to do with clutch size, but is caused by parental reaction to the disturbance to the nest that occurred when the extra egg was added to the treatment nests by the experimenter. To avoid this criticism, a better control manipulation might have been to do exactly the same thing to control nests as treatment nests, except that as soon as the alien egg is added, it was immediately removed again by the experimenter.

Another example might be in investigating the effect of a pacemaker on the longevity of laboratory rats. Now, applying a surgical procedure to a rat is likely to be stressful, and this stress may have effects on longevity that are nothing to do with the pacemaker. You might consider subjecting the control group to the same surgical procedures as the treatment group with the sole exception that the pacemaker is not inserted, or is inserted then immediately removed, or a non-functional pacemaker is inserted. This controls for any confounding effects due to the anaesthetic or other aspects of the surgical procedure. Although this type of control has attractions, ethical considerations may count against it (see Section 2.3.3).

You might feel a bit of a fool going through what seem like elaborate charades (or *sham procedures* as the jargon would have it). You should not; the fool is the person who does not strive for the best possible control.

One last thing: make sure that all manipulations and measurements on the treatment and control experimental units are carried out in a random order. If you measure all the manipulated units, then all the controls, you introduce 'time of measurement' as a potential confounding factor (see Section 11.1).

Choosing the best control group for your experiment warrants careful thought.

2.3.3 The ethics of controlling

Sometimes a negative control may be unethical. In medical or veterinary work, deliberately withholding treatment to those in need would be hard to justify. In the example earlier involving surgical procedures on rats, ethical considerations may cause you to reject the type of control suggested in favour of one that involved less animal suffering. For example, if you could argue (based on previous work) that the surgical procedures themselves (rather than the pacemaker) were highly unlikely to affect longevity, then a control group of unmanipulated animals might be the best compromise between scientific rigour and ethical considerations. Further, if we were dealing with a species where even keeping the animals in captivity was stressful, then we might even consider using a historical control (if one were available) in order to avoid having to keep a control group. But such choices must be made extremely carefully. While the use of historical controls may reduce the suffering of animals in the experiment—an aim which we cannot emphasize too strongly—if the controls are inappropriate and will not convince others, then all the animals in the experiment will have suffered for nothing and we would have been better not doing the experiment at all.

See Section 2.3.3 of the supplementary material for more discussion on controls, especially on the ethics of controls, the placebo effect, and case-control studies. Go to: **www.oxfordtextbooks.co.uk/ orc/ruxton4e/**.

 Be careful you don't adopt a poor historical control for ethical reasons; better you do no experiment at all than a poor one.

2.3.4 Situations where a control is not required

In view of what we have discussed so far, you might be surprised at our suggestion that sometimes a control is not necessary. To see why this might be the case, consider answering the following research question:

Which of three varieties of carrot grows best in particular conditions?

Here there is no need for a control, as we are only comparing growth rates between the three treatment groups—we are not comparing them with some other reference growth rate. Similarly, there is no need for a control in an experiment to address the following:

How does the regularity of feeding with a liquid fertilizer affect the growth of tomato plants?

Here, we again want to compare different treatment groups (i.e. groups of plants fed at different rates), and there is no need for a separate control. However, the selection of appropriate treatment groups to answer such questions is an art in itself (see Box 6.3).

Sometimes, but not often, a control group is not required. A clearly stated hypothesis will help you spot such cases.

2.4 The importance of a pilot study and preliminary data

It seems inevitable that as soon as you begin a scientific study you will find yourself under time pressure. Undergraduate projects are generally only brief, a PhD never seems to last long enough, and the end of a post-doctoral grant begins looming on the day that it starts. Such pressures can lead to a temptation to jump straight in and do some experiments quickly. Having reams of data can be very comforting. However, the old adage of 'more haste; less speed' applies to biological studies as much as anything, and an important step in experimental biology that is often missed is spending a bit of time at the beginning of a study collecting some pilot observations. A pilot study can mean anything from going to the study site and watching for a few hours or days, to trying a run of your experiment on a limited scale. Exactly what it entails will depend on the particular situation. Leaping straight into a full-scale study without carrying out such pilot work will often mean that you get less from the study than you might have done if the pilot study had suggested some fine-tuning. Not infrequently, your three-month-long full-scale study will be totally useless because of a pitfall that a three-day pilot study would have alerted you to. In Section 2.1 we suggested that a pilot study can be invaluable in helping you to develop interesting, focused questions. In Sections 2.4.1 and 2.4.2 we will outline several other advantages of pilot studies.

2.4.1 Making sure that you are asking a sensible question

One aim of a pilot study is to allow you to become well acquainted with the system that you will be working on. During this phase of a project, you will be trying to gain information that will help you to better design your main experiments. But what should you be aiming to do in this time? Probably the most important goal of a pilot study is to validate the biological question that you are hoping to address. You may have read about stone-stealing behaviour in chinstrap penguins, a phenomenon where penguins steal stones from each others' nests and incorporate them into their own nest. Let's say that you decide that this would make an interesting subject for your investigation. Fortunately, there is a chinstrap colony at the local zoo that you will be able to work on. So you form hypotheses about aspects of stone stealing, and design an elaborate experimental programme to evaluate these hypotheses. This involves placing all sorts of manipulations on penguins' nests, to add and remove stones of different sizes and colours. Setting up the experiment is hard work, and the penguins suffer considerable disturbance. However, eventually it is done and you sit back in your deckchair to begin collecting the data. After a day, you have seen no stone-stealing behaviour, but you are not bothered: stone stealing is a well-documented phenomenon, and there are plenty of papers written on the subject. Unfortunately, your particular penguins haven't read these papers and after a month of observing and not seeing a single case of stone stealing you give up.

If only you had taken the time to watch the penguins for a couple of days before putting in all your hard work. You would have realized that stone stealing was a rare occurrence in this colony, and that you might have to rethink your project. You might also have seen some other interesting behaviour that could have formed the basis of a valuable alternative study. Alternatively you might have observed two very different types of stealing behaviour, allowing you to plan your project in such a way that you can look at these two different types, making for an even better project.

It is worth mentioning at this point the dangers of relying on others to give you this information, no matter how august they are, unless you absolutely have to. Your supervisor might tell you that stone stealing will occur in your study colony. Maybe they are basing this on having read the same papers that you read; or on a discussion with a zoo keeper, who on the basis of having seen a case two years ago proudly asserts that stone stealing goes on at their colony. At the very least you should enquire when the behaviour was seen and how often. In the end, it is your project, you will be the one putting in the hard work and wasting time if it goes wrong; so you should make the preliminary observations that convince you that your main study is possible. Some things you have to take on trust, but don't take things on trust if you can easily check them for yourself.

It may be that you consider our unfortunate penguin-watcher to be a straw man, and that there is no way that you would be so foolish. We can only say that we have encountered a large number of analogous abortive experiments and have been the guilty party ourselves on enough occasions to make us squirm.

 Pilot studies save the potentially considerable embarrassment, wasted expense, and perhaps animal suffering of unsuccessful experiments.

2.4.2 Making sure that your techniques work

A second crucial role for a pilot study is that it gives you a chance to practise and validate the techniques that you will use in the full study. This might involve something as simple as making sure you will be able to make all the observations on the penguins that you need from the comfort of your deckchair, or that you won't disturb the animals too much. You might also decide to check that you can use the shorthand code that you have devised to allow you to follow the behaviour of the birds, and that you have a code for all the behaviours you need. Or you might need to check that you can take all the measurements you need in the time you have available. Of course, this doesn't just apply to field projects. You may have planned an experiment that requires you to count bacteria with a microscope, but how long can you spend staring down a microscope before your accuracy deteriorates? The only way to know is to try it, and this will ultimately affect the maximum size of the experiment you can plan for and carry out.

There is a lot to be said for practising all the techniques and methods that you will use in your experiment before you use them in your main experiment. If your experiment will involve dissecting mosquitoes, you want to make sure you can do this efficiently, so as to be sure that you will not destroy valuable data points and cause unnecessary animal suffering.

Trial runs of different types of assays can also allow you to fine-tune them so that they will give you useful results. You might be planning a choice test for female butterflies in which you will give a female a choice of several host plants and see which she lays her eggs on. If you leave her for too long to lay eggs, she may lay eggs on all the hosts, and subtle preferences will be hard to detect. Conversely, if you leave her for too little time, she may not have sufficient time to properly explore all the hosts, so may lay eggs at random. By carrying out pilot experiments with females left for different lengths of time, you can work out what time interval to use in the main experiment.

Pilot studies also provide an opportunity for techniques to be standardized in studies involving several observers. Perhaps you need to check that what one person regards

---

STATISTICS BOX 2.1 Obtaining basic data to fine-tune design and statistics

An advantage of actually carrying out a pilot study in the form of a small-scale version of the actual experiment that you have planned is that if your techniques work well, you will have some preliminary experimental data of the sort that you will collect in the full study. This can be extremely valuable in two ways.

Firstly, such data can provide the necessary information to see whether the planned size of your experiment is likely to be appropriate. This will be discussed in more detail when we describe statistical power in Chapter 6.

Secondly, pilot data can give you something to try out your planned statistical tests on. Whenever you carry out a study you should have planned how you will analyse the data that you collect, and the best way to decide whether your analysis will work is to try it on data from a pilot experiment. Every statistical test requires data of a specific type that must meet a number of assumptions, and there is nothing worse than finding at the end of a study that you have collected your data in a way that makes it impossible to analyse (or much more difficult than was necessary). Such problems will immediately become apparent if you try to analyse your pilot data, allowing you to modify the way you do your study before it is too late. This pilot analysis also has yet another advantage. Like most things, manipulating data and performing statistical tests gets easier with practice. Once you have learnt how to do the appropriate specific analysis on your pilot data, it will then be easy to do the same analysis on the full data set. It will almost inevitably be the case that your data collection takes slightly longer to do than you had planned, leaving you less time at the end of the study to analyse the data and write the results up. If this is the point when you first worry about statistics, this is likely to put you under even more time pressure. If, on the other hand, you have already dealt with all the details of the statistics early in the study, the analysis will take much less effort at the end, allowing you to spend more time writing and thinking about what your results mean. Indeed, even in those situations where you can't actually obtain pilot data yourself, we would encourage you to either borrow some data from someone else, or make up a dummy data set of the sort that you expect to collect, in order to try your analysis out. With statistics, as with all the other techniques you use, the best way to ensure they will work when they need to is to try them first when the pressure is off.

Q 2.5 The research task you've been assigned is to observe whether flock size in wild geese grazing on agricultural land varies with species and with land use. What aspects of this study would you seek to evaluate in a pilot study before you begin?

as an aggressive behaviour is the same for all observers, or that all are measuring things in the same way or from the same positions. See Section 11.5 for more on this. Another advantage of a pilot study is that it can provide a small-scale data set, of the same form as your main planned experiment will produce. This gives you a chance to test your planned statistical analysis: see Statistics Box 2.1 for more on this.

In short, many of the problems that might occur in the main experiment can be ironed out by careful thought beforehand. However, nothing really brings out all the possible flaws and problems with an experiment like trying out the techniques under the conditions in which you will use them. We encourage you to do this whenever possible.

If you would like to test your understanding of the material in this chapter a little more, then we have provided some further self-test questions in the supplementary material. But if you feel that you have mastered the concepts covered in this chapter, we can move on in Chapter 3 to consider what broad type of study might be more effective for addressing a given set of hypotheses.

 Practise all your data collection techniques in a realistic setting before you need to use them for real.

Summary

- You cannot design a good experiment unless you have a clear scientific hypothesis that you can test.
- Pilot studies help you form good hypotheses, and test your techniques for data collection and analysis.
- Make sure that your experiment allows you to give the clearest and strongest evidence for or against the hypothesis. Careful selection of control groups is often key to this.
- Make sure that you can interpret all possible outcomes of your experiment.
- Use indirect measures with care.
- Design experiments that give definitive answers that would convince even those not predisposed to believe that result.

3 Selecting the broad design of your study

- A key decision in your design is likely to be whether to perform an experimental manipulation or simply record natural variation. Both techniques have their advantages and disadvantages; and a combination of the two is often effective (Section 3.1).
- You may then have to address whether to study your subjects in their natural environment or in the more controlled conditions of the laboratory (Section 3.2) or whether to carry out experiments *in vitro* or *in vivo* (Section 3.3).
- All these decisions involve an element of compromise, and whilst there are good and bad scientific studies (and all the shades in between), there are no perfect ones (Section 3.4).

3.1 Experimental manipulation versus natural variation

Once you have chosen your research question, you then face the problem of how you are going to test your specific hypotheses. This is the crux of experimental design. One of the major decisions you will have to make is whether your study will be **manipulative** or **correlational**. A manipulative study, as the name suggests, is where the investigator actually does something to the study system and then measures the effects these manipulations have on the things that they are interested in. In contrast, a correlational study makes use of naturally occurring variation, rather than artificially creating variation, to look for the effect of one factor on another. A correlational study may also be called a *mensurative experiment* or an **observational** study.

> In a **manipulative** study the experimenter changes something about the experimental system then studies the effect of this change, whereas **observational** studies do not alter the experimental system.

3.1.1 An example hypothesis that could be tackled by either manipulation or correlation

Imagine that we had a hypothesis:

> Long tail streamers seen in many species of bird have evolved to make males more attractive to females.

From this we could make the straightforward prediction that males with long tails should get more matings than males with short tails, but how would we test this?

Correlational approach

One way to test the prediction would be to find ourselves a study site and catch a number of males. For each male that we catch, we measure the length of their tail streamers, add leg bands to the birds to allow us to distinguish individuals, and then let them go again. We could then watch the birds for the rest of the season and see how many females mate with each male. If, after doing the appropriate statistical analysis, we found that the males with longer tails obtained more matings, this would support our hypothesis. A study like this is a correlational study—we have not actually manipulated the factor we are interested in ourselves, but made use of the naturally occurring variation in tail length to look for a relationship between tail length and number of mates. It is worth mentioning that some people get confused by the term correlational study, assuming that the name implies that the statistical test of the same name must be used to analyse the data collected. It does not.

Manipulative approach

An alternative approach would be to actually manipulate the length of the tail. So again, we would catch the male birds, but instead of just measuring them and letting them go, we would manipulate their tail length. For example we might assign the birds into three groups. Birds in group 1 have the ends of their tail streamers cut off and stuck back on again so that their streamer length is unchanged; these would form the control group. Birds in group 2 would have the ends of their tail streamers cut off, reducing their length. Birds in group 3 would have their tails increased in length by first cutting off the end of their streamers then sticking them back on, along with the bits of streamers that were removed from the birds in group 2. We then have three groups all of which have experienced similar procedures, but one resulted in no change in streamer length, the next involves shortening of the streamers and the last involves elongating them (see Figure 3.1). We would then add the leg bands, let the birds go, and monitor the numbers of matings. If the birds in the group with the extended tails got more matings than the birds in the other groups, this would support our hypothesis. If the birds that had had their streamers reduced also got fewer matings than the controls with unchanged length, this would also add further support for our hypothesis.

In this example, both the manipulative and correlational techniques seem effective; next we discuss the factors that must be considered to decide which would be most effective.

 Both manipulative and correlative studies can be effective, and the best approach will depend on the specifics of your situation.

3.1.2 Arguments for doing a correlational study

Correlational studies have several advantages. They are often easier to carry out than manipulative studies. Even from the number of words that we've just used to describe

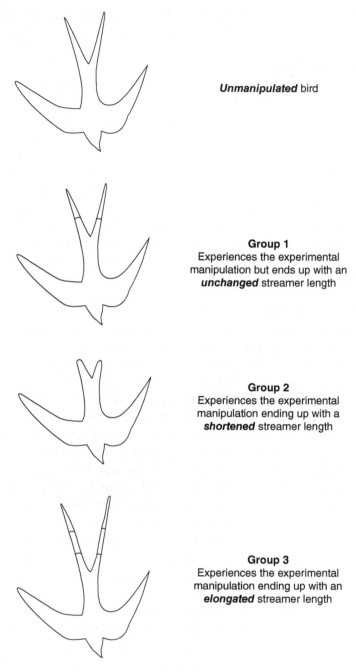

Unmanipulated bird

Group 1
Experiences the experimental
manipulation but ends up with an
unchanged streamer length

Group 2
Experiences the experimental
manipulation ending up with a
shortened streamer length

Group 3
Experiences the experimental
manipulation ending up with an
elongated streamer length

Figure 3.1 For the experiment exploring the relationship between streamer length and mating success in a bird species discussed in Section 3.1.1, we argue that three experimental groups are required. Each group experiences manipulation of its streamers, but this manipulation has a different effect on streamer length in the three cases: in Group 1 the streamers are unchanged in length after manipulation, in Group 2 they are shortened, and in Group 3 they are increased in length.

these two different ways of testing our hypothesis about tail streamers, it is clear that the correlational study will often involve less work. This may not just be an advantage through saved time and effort, it may also mean that you have to handle or confine organisms for much less time, if at all. If we are dealing with organisms that are likely to be stressed or damaged by handling, or samples that can be contaminated, this is obviously a good thing.

A major worry of doing manipulative studies is that the manipulation that you have carried out will have unintended effects. Biological organisms are integrated units and any change in one part may have profound effects on other functions. If we cut off part of a male's tail, this might affect his ability to fly as well as his attractiveness, and any results we see as a consequence might be due to changes in his flight ability, and not due to his tail length directly. Even in fields where modern techniques of genetic engineering allow single genes to be modified or knocked out, it is impossible to be sure that our manipulation doesn't alter things other than the characters we are interested in. If we design our experiments carefully, with adequate and appropriate controls, we should be able to detect such spurious effects if they occur, but we should always be cautious. In the current study, we could imagine how all our manipulations might impair flight performance and so foraging ability, thus causing our study individuals to be in poor condition. The overall lower condition of birds in our sample would mean that our ability to generalize our conclusions to birds with a range of body conditions is weakened. Further, the poor condition of all our birds might reduce their attractiveness to females, making any difference between groups more difficult to detect (a floor effect: see Section 11.6.5). Such problems will not arise in a correlational study.

A final benefit of correlational studies is that we can be sure that we are dealing with biologically relevant variation, because we haven't altered it. Suppose that we extend or reduce tail streamers by 20 cm, but in nature, tail streamers never vary by more than 2 cm across the whole population. Since we have modified birds such that they have traits that are far outside the naturally occurring range, it is doubtful that our experiment can tell us anything biologically relevant about the system. Again, this need not be a fatal problem of manipulative studies. As long as manipulations are carefully planned and based on adequate biological knowledge and pilot data of the system, this problem can be avoided. That said, we urge you to think carefully about the biological relevance of any manipulation that you plan before you carry it out.

 Correlational studies are generally easier to perform, and have less chance than manipulative studies of going badly wrong.

3.1.3 Arguments for doing a manipulative study

Given all the issues raised in the preceding section, you might now be wondering why anyone ever bothers to do a manipulative study at all. Despite the advantages offered

A **third variable** (also called a *confounding factor* or *confounding variable*) is a variable (*C*) that separately affects the two variables that we are interested in (*A* and *B*), and can cause us to mistakenly conclude that there is a direct link between *A* and *B* when there is no such link.

by correlational studies they suffer from two problems: **third variables** and *reverse causation*. As problems go, these can be big ones.

Third variables

We can sometimes mistakenly deduce a link between factor *A* and factor *B* when there is no direct link between them. This can occur if another factor *C* independently affects both *A* and *B*. *C* is the third variable that can cause us to see a relationship between *A* and *B* despite there being no mechanism providing a direct link between them.

As an example, imagine that we survey patients in a general practitioner's (GP's) waiting room and find that those that travelled to the GP by bus had more severe symptoms than those who travelled by car. We would be rash to conclude from this that travelling by bus is bad for your health. The cause of the correlation is likely to reside in a third variable: for example, socioeconomic group. It may be that there is no direct link between bus travel and poor health, but those from lower socioeconomic groups are likely to be both in poorer health and (separately) more likely to use buses than the more affluent.

Let's go back to the correlational study earlier where we found a relationship between tail streamer length and the number of matings that a male obtained. Does this show that the length of a male's tail affects the number of matings that he gets? What if females actually base their choice on the territory quality of a male, not on his tail length at all, but males on good territories are able to grow longer tails (maybe they can get more food on the better territories). We might think that we have observed a causal relationship between the number of matings that a male gets and his tail length, but what are actually occurring are separate relationships between the quality of territory and both the number of matings that the male gets and his tail length. Territory quality is a third variable that affects both male tail length and the number of mates that the male gets. Because it affects both, it leads us to see a relationship between the number of matings that a male gets and the length of his tail, even though females do not consider tail length when evaluating a male at all. In almost any correlational study there will be potential third variables that we have not measured and that could be producing the relationship that we observe. It is important that you understand the concept of third variables clearly, so let's look at some further examples.

Ronald Fisher, a famous (and sometimes eccentric) evolutionary biologist, statistician, and pipe-smoker, is reputed to have used third variables to discount the observed relationship between smoking and respiratory diseases. Smoking, he argued, does not cause diseases; it is just that the types of people that smoke also happen to be the types of people that get diseases for other reasons. Maybe these are diseases caused by stress, and stressed people are also more likely to smoke.

More women graduates never marry than would be expected for the population as a whole. This does not mean that going to university makes a woman less likely to marry as such (or that no one would want to marry the sort of male you meet at university). It perhaps means that personality traits or social circumstances associated with a slightly increased chance of going to university are also associated with a slightly decreased chance of marrying.

There is a correlation between the amount of ice cream sold in England in a given week and the number of drowning accidents. This is unlikely to be a direct effect: that, say, heavy consumption of ice cream leads to muscle cramp or loss of consciousness among swimmers. More likely a third variable is at work: when the weather is warm then more ice creams are sold but also (and separately) inexperienced sailors hire boats and people who are not strong swimmers dive into rivers and lakes.

In short, third variables mean that it is difficult to be confident that the pattern we have observed in an unmanipulated system is really due to the factors that we have measured and not due to correlation with some other unmeasured variable. The only way to be certain of removing problems with third variables is to carry out experimental manipulations.

Reverse causation

The second problem of correlational studies is **reverse causation**. This can occur when we see a relationship between factors A and B. It means mistakenly assuming that factor A influences factor B, when in fact it is change in B that drives change in A.

For example, imagine a survey shows that those who consider themselves to regularly use recreational drugs also consider themselves to have financial worries. It might be tempting to conclude that a drug habit is likely to cause financial problems. The reverse causation explanation is that people who have financial problems are more likely than average to turn to drugs (perhaps as a way to temporarily escape their financial worries). In this case, we'd consider the first explanation to be the more likely; but the reverse causation explanation is at least plausible. Indeed, both mechanisms could be operating simultaneously.

Let's consider another example of reverse causation. There is a correlation between the number of storks nesting on the chimneys of a Dutch farmhouse and the number of children in the family living in the house. This sounds bizarre until you think that larger families tend to live in larger houses with more chimneys available as nest sites for the storks. Large families lead to more opportunities for nesting storks; storks don't lead to large families!

Reverse causation is not a universal problem of correlational studies. Sometimes reverse causation can be discounted as a plausible explanation. In our correlational study of male tail streamer length and mating success described earlier, it would be extremely difficult to imagine how the number of matings a male gets later in a season could affect his tail length when we catch him at the beginning of the season. However, imagine that we had instead caught and measured the males at the end of the season. In that case it might be quite plausible that the number of matings a male gets affects his hormone levels. The change in hormone levels might then affect the way that the male's tail develops during the season in such a way that his tail length comes to depend on the number of matings that he gets. In that case we would observe the relationship between tail length and matings that we expect, but it would be nothing to do with the reason that we believe (females preferring long tails). Such a pathway might seem unlikely to you and you may know of no example of such a thing happening. However, you can be sure that the Devil's Advocate will find such arguments highly compelling unless

Q 3.1 In a survey of road-killed badgers it was found that those with higher levels of intestinal parasites had lower bodyweights: can we safely conclude that intestinal parasites cause weight loss or is there a plausible third variable effect that could explain this observation?

Reverse causation is mistakenly concluding that variable A influences variable B when actually it is B that influences A.

Q 3.2 Would you opt for correlation or manipulation if you wanted to explore the following:

(a) Whether having a diet high in processed foods increased a person's propensity for depression.

(b) Whether a diet high in saturated fats increases the risk of a heart attack.

you can provide good evidence to the contrary. Experimental manipulation gets around the problems of reverse causation, because we manipulate the experimental variable.

These two problems of non-manipulative studies are captured in the warning, 'correlation does not imply causation' and are summarized in Figure 3.2. Many of the hypotheses that we will find ourselves testing as life scientists will be of the general form 'some factor A is affected by some factor B' and our task will be to find out if this is indeed true. If we carry out a correlational study and find a relationship between A and B, this provides some support for our hypothesis. However, the support is not in itself conclusive. Maybe the reality is the A affects B through reverse causation or that some other factor (C) affects both A and B, even though A and B have no direct effect on each other. Such problems mean that at best our correlative study can provide data that are *consistent* with our hypothesis. If we wish to provide unequivocal evidence that A affects B, then we must carry out a manipulative study. Thus, we would encourage you to think carefully before embarking on a solely correlational study.

→ Third variable (and to a lesser extent) reverse causation effects are big drawbacks of correlational studies that are likely to make a manipulative study more attractive.

3.1.4 Situations where manipulation is impossible

In the last section we were pretty harsh about correlational studies. Yet many correlational studies are carried out every day, and the scientific journals are full of examples. There are a number of situations in which correlational studies can provide very valuable information.

Sometimes it is simply not possible to manipulate. This may be for practical or ethical reasons. You can imagine the (quite reasonable) outcry if, in an attempt to determine the risks of passive smoking to children, scientists placed a number of babies in incubators into which cigarette smoke was pumped for eight hours a day. Such problems are common in studies relating to human health and disease (epidemiology) and, as a consequence, correlational studies fulfil a particularly important role in this field (we discuss aspects of their use in this field in Box 3.1).

It may also not be technically feasible to carry out the manipulation that we require in order to test our particular hypothesis. Modifying streamer length was a difficult enough manipulation; experimentally modifying beak length or eye colour in a way that does not adversely affect the normal functioning of subject birds would be nearly impossible.

A second valuable role for correlational studies is as a first step in a more detailed study. Imagine that you want to know what sorts of factors affect the diversity of plants in a given area. One possible approach would be to make up a list of everything that you think might be important—all the soil chemicals and climactic factors—and then design a programme of experiments to systematically examine these separately and then in combination. Such a study would be fine if you had unlimited research funds (not to mention unlimited enthusiasm and research assistants). However, a more efficient approach might be to begin with a large correlational study to see which factors seem to

We see a relationship between A and B

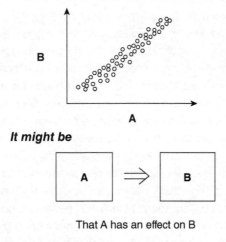

It might be

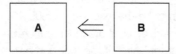

That A has an effect on B

But it could be

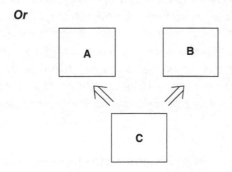

That cause and effect are the other way around and B
has an effect on A. (This is reverse causation.)

Or

That A and B have absolutely no effect on each other,
but both are affected by a third variable C

Figure 3.2 When we find a relationship between two variables (*A* and *B*) then
we must be careful in how we interpret the mechanism underlying the observed
relationship, since a number of (non-exclusive) possibilities exist. If we want to
argue that the likely explanation is that the value of *A* influences the value of *B*,
then we need to demonstrate that this is a more plausible mechanism than two
alternatives: reverse causation and third variable effects. In our case, the reverse
causation mechanism is that variable *B* strongly influences *A* (rather than vice
versa); the third variable explanation is that actually there is no direct mechanism
linking *A* and *B*, but both *A* and *B* are strongly linked to a third variable (*C*).

BOX 3.1 The role of correlative studies in human epidemiology

Correlation studies are particularly important in epidemiology: the study of the factors relating to health and illness in a population, where it is generally neither practical nor ethical to carry out manipulations. Ethically, manipulations generally have to be changes that are expected to have a neutral or positive impact on the health and welfare of the subjects involved. Thus the most common use of manipulative study in the medical sciences is in clinical trials of new therapies. Such studies generally take a great deal of organization to set up because it can be very easy for conflict to develop between interventions that are perceived to be best for an individual patient and the needs of the trial. If during the trial, having been randomized to a new treatment, a patient does not seem to be responding as well to treatment as the doctor would expect, then there is an understandable tendency for the doctor to want to switch that patient to the established treatment. Conversely, if over the course of a trial it becomes clearer that a novel treatment is performing much better or worse than the established treatment, then doctors involved in the study might (again reasonably) want to call the trial to a halt and treat all their patients in what appears to them to be the most efficacious manner. These problems can be resolved to some extent by blind procedures (see Section 11.6.3), by briefing participants clearly about the value of the trial, and by having protocols in place to stop the trial earlier than planned if interim analyses of partial results provide a clear conclusion. Informed consent is a normal requirement for patients to take part in a clinical trial, and this can be time-consuming to achieve and needs to be done very carefully to avoid bias arising from self-selection of subjects. Getting patient agreement can be particularly challenging where placebos are involved. One large-scale trial of a potential AIDS treatment was greatly damaged by informal groups of patients enrolled in the trial pooling their drugs such that even those assigned to the placebo would get some of the potential benefit from the new drug. For these reasons correlational studies are particularly valuable in the medical sciences.

be most important. Once potentially influential factors had been selected, manipulative studies could be used in a more targeted way to confirm and refine these findings.

So ultimately there is nothing wrong with correlational studies—they fulfil an important role in the life sciences. Often a correlational study followed by carefully targeted manipulative studies will provide more compelling arguments than either would have provided alone. Correlational studies can have weaknesses, and the conclusions that we can draw will ultimately be limited by these weaknesses. However, as long as we are aware of these weaknesses, and do not overstate our results, correlational studies can be very useful.

 Correlational studies can be a useful and practical way to address biological questions, but their limitations must always be borne in mind.

3.2 **Deciding whether to work in the field or the laboratory**

Another decision that you will often be faced with in many life science studies is whether it will be laboratory-based or field-based. Here the answer will depend on the question you are asking, and the biology of the system concerned. There are advantages and disadvantages to both approaches. Suppose that we are interested in knowing whether provisioning a larger clutch of young reduces a female zebra finch's survival. We decide that we will manipulate clutch size, by removing or adding eggs to zebra finch nests, and then monitoring the survival of the mother after the chicks have fledged.

We could potentially carry out such an experiment in either the lab or the field. How should we choose? Let's begin with the lab. Maybe the most obvious question to ask is whether your study organism will be comfortable in the lab. In this case, we are unlikely to have a problem, as zebra finches will breed readily in captivity; but if we had been looking at the same question in a bird that is more challenging to keep in captivity (like an albatross), we might have had problems. The suitability of an organism for lab studies is hugely variable, and must be given careful thought. If an animal is not well disposed to captivity, then there are obvious welfare implications as well as the risk that your results will have limited relevance to the situation in nature.

Now, assuming that your organism will be happy in the lab, there are several advantages of a lab study. In general it will be far easier to control conditions in a lab setting, allowing you to focus on the factors of interest without lots of variation due to other uncontrolled factors. Things like temperature, daylight, and food supply can all have profound effects on organisms, and variation due to these factors might mask the effects we are interested in. We can also ensure that all of our animals are well fed or our plants well watered, removing between-individual variation that could be caused by these factors in a natural setting. Observation will also generally be much easier in the lab. While the thought of watching zebra finches in the sunny Australian outback may sound idyllic, the reality of spending weeks trying to find your study birds is far less attractive. It will nearly always be easier to make detailed measurements on most organisms in the lab.

On the other hand, the controlled nature of the lab environment is also its major drawback. Laboratory individuals will not experience many of the stresses of everyday life in the field, and that can make it difficult to extrapolate from lab results. Suppose that we find absolutely no effect of our laboratory manipulations on the lifespan of female zebra finches. What can we conclude? We can conclude that the manipulation had no measurable effect *under the particular laboratory conditions that we used*. This is very different from saying that it would have had no effect in the wild, which is presumably what we are really interested in. It is quite conceivable that our well-fed laboratory birds in their temperature-controlled rooms, free from parasites, can easily cope with extra chicks, but that in the harsher environment of the field they would have suffered a large cost. Of course, if we found no effect in the field, this result would only apply to the specific environmental conditions that occurred during our study; but those conditions are still likely to be more representative than the conditions in a lab. Thus, it is usually safer to generalize from a field study than a lab study. Generalization is a subject that is at the heart of statistical methodology, so we explore it a little further in Statistics Box 3.1.

Sometimes, our research question means that a lab study is impractical. If we are interested in differences between the sexes in effort expended in feeding the young in the nest using zebra finches, then it would be very difficult to create foraging conditions in an aviary where the conditions were close enough to those naturally experienced by wild birds. In the wild parents will bring back a great diversity of food types gathered from a wide variety of microhabitats sometimes hundreds of metres distant. This would be impossible to faithfully recreate in the lab. The danger is that any aviary experiments that you conducted on this would be discounted by Devil's Advocates as being entirely artefactual, due to your asking birds to operate in a foraging situation far removed from that which they have been evolutionarily adapted to cope with. However, if your research question is about how egg laying affects the distribution of reserves of calcium in different parts of females' bodies, then it would be easier to argue that this is less likely to be influenced by the environmental conditions, and hence a laboratory experiment might be reasonable. So, as ever, *the most appropriate approach depends on your research question.*

STATISTICS BOX 3.1 Generalizing from a study

An important issue in the interpretation of a study is how widely applicable the results of the study are likely to be. That is, how safely can you generalize from the particular set of subjects used in your study to the wider world? Imagine that we measured representative samples of UK adult males and females and found that males in our sample were on average 5 cm taller. Neither we nor anyone else are particularly interested in our specific individuals, rather we are interested in what we can conclude more widely from study of the sample. This becomes easy if we think about the population from which our sample was drawn. Our aim was to create a representative sample of UK adults. If we have done this well, then we should feel confident about generalizing our results to this wider population. Can we generalize even wider than this: say to European adults or human adults in general? We would urge a great deal of caution here. If your interest was in European adults, then you should have designed your study in the first place to obtain a representative sample of European rather than only UK adults. However, providing you add an appropriate amount of caution to your interpretation then it may be appropriate to use your knowledge of biology to explore how your results can be extrapolated more widely than your original target population. In this specific case, the specific difference of 5 cm would probably have little relevance beyond your specific target population of UK adults, since other countries will differ significantly in factors (e.g. racial mix, diet) that are highly likely to affect height.

Our advice would be to aim to work in the field—generally giving results that are easier to generalize from—unless it is impracticable to do so.

3.3 *In vivo* versus *in vitro* studies

In many ways, the question of lab or field has analogies in the biomedical sciences with whether to carry out studies *in vivo* or *in vitro*. If we want to know whether an antimalarial drug alters the reproductive strategy of a malaria parasite, do we measure this using parasites in Petri dishes or within organisms? *In vitro* will generally be easier to do (although this is certainly not always the case), easier to control, and easier to measure. But do the results have any bearing on what would happen in the natural system? The most effective approach will depend on the details of your study and there are advantages and disadvantages to each. Indeed the best study would probably contain evidence from both approaches.

 Issues of generalization of results will lead you towards *in vivo* experiments, whereas issues of practicality will lead you to *in vitro*.

3.4 **There is no perfect study**

Hopefully by now you will have got the impression that there is no perfect design that applies to all experiments. Instead, the best way to carry out one study will be very different from the best way to carry out another, and choosing the best experimental approach to test your chosen hypothesis will require an appreciation of both the biology of the system and the advantages and disadvantages to different types of studies. This is why *biological insight is a vital part of experimental design*. Let's illustrate these points by considering another example: the relationship between smoking and cancers.

In a sense, all the data that we collect tells us something, but whether this something is useful and interesting can vary dramatically between data sets. Some data will be consistent with a given hypothesis, but will also be consistent with many other hypotheses too. This is typical of purely correlational studies, where it is difficult to discount the effects of reverse causation or third variables. A positive relationship between lung cancer and smoking definitely provides support for the idea that smoking increases the risk of lung cancer, but it is also consistent with many other hypotheses. Maybe it is due to stress levels that just happen to correlate with both smoking and cancer. Or maybe stress is an important factor, but there is also a direct link between smoking and cancer. In order to draw firm conclusions, we need something more. Our confidence might be improved by measuring some of the most likely third variables—like stress or socio-economic group—to allow us to discount or control for their effects. This would reduce the possibility that an observed relationship was due to some other factor, but would not remove it entirely (since there may still be some confounding factor that we've failed to account for).

If we were able to do an experiment where people were kept in a lab and were made to smoke or not smoke, any relationship between cancer and smoking emerging from such a study would be far more convincing evidence that smoking caused cancer. If this could be repeated on people outside the lab, the evidence that smoking caused cancer in a real world setting would become compelling. However, such experiments are ethically intolerable. Even if we can't do these experiments we can do other things to increase our confidence. Maybe we can demonstrate a mechanism. If we could show in a Petri dish that the chemicals in cigarette smoke could cause cells to develop some pathology, this would increase our confidence that the correlational evidence supported the hypothesis that smoking increases the risk of cancer. Similarly, experiments on animals might increase our confidence. However, such experiments would entail a whole range of ethical questions, and the generality of the results to humans could be questioned.

Good experimental design is all about maximizing the amount of information that we can get, given the resources that we have available. Sometimes the best that we can do will be to produce data that provide weak evidence for our hypothesis. If that is the limit of our system, then we have to live with this limitation. However, if what we can conclude has been limited not by how the natural world is but by our poor design, then we have wasted our time and probably someone's money too. More importantly, if our experiment has involved animals these will have suffered for nothing.

Hopefully, if you think carefully about the points in this chapter before carrying out your study, you should avoid many of the pitfalls. If you want to test your knowledge of the concepts introduced in this chapter a little more, then we offer more self-test questions in the supplementary information. Otherwise, let's proceed to the next chapter, where we explore how we meet one of the central challenges in experimental design: dealing with the variation between individuals that is widespread in biology.

 There are always compromises involved in designing an experiment, but you must strive to get the best compromise you can.

Summary

- Correlational studies have the attraction of simplicity but suffer from problems involving third variable effects and reverse causation.
- Manipulative studies avoid the problems of correlational studies but can be more complex, and sometimes impossible or unethical.
- Be careful in manipulative studies that your manipulation is biologically realistic and does not affect factors other than the ones that you intend.
- Often a combination of a correlational study followed by manipulation is very effective.

- The decision whether to do animal experiments in the field or laboratory will be influenced by whether the test organism can reasonably be kept in the laboratory, whether the measurements that need to be taken can be recorded in the field, and on how reasonably laboratory results can be extrapolated to the natural world. All these vary from experiment to experiment.

- There is no perfect study, but a little care can produce a good one instead of a bad one.

4

Between-individual variation, replication, and sampling

- In any experiment, your experimental subjects will differ from one another.
- Experimental design is about removing, or controlling for, variation due to factors that we are not interested in (random variation or noise), so we can see the effects of those factors that do interest us (Section 4.1).
- One key aspect to coping with variation is to measure a number of different experimental subjects rather than just a single individual, in other words to replicate (Section 4.2).
- Having identified a population of independent experimental subjects, we then need to sample from that population—and this can be done in a number of different ways (Section 4.3).

4.1 Between-individual variation

If we are interested in how one characteristic of our experimental subjects (variable *A*) is affected by two other characteristics (variables *B* and *C*), then *A* is often called the *response variable* or *dependent variable* and *B* and *C* are called *independent variables*, *independent factors*, or sometimes just **factors**. It is likely that *A* will be affected by more than just the two factors *B* and *C* that we are interested in.

Any variation in the response variable between individuals in our sample (between-individual variation) that cannot be attributed to the independent factors is called **random variation**, *inherent variation*, *background variation*, *extraneous variation*, *within-treatment variation*, or **noise**.

Wherever we look in the natural world, we see variation: salmon in a stream vary in their body size; bacteria in test tubes vary in their growth rates. In the life sciences, more than physics and chemistry, variation is the rule; and the causes of variation are many and diverse. Some causes will be of interest to us: bacteria may vary in their growth rates because they are from different species or because a researcher has inserted a gene into some that allows them to utilize an additional type of sugar from their growth media. Other causes of variation are not of interest: maybe the differences in growth rates are due to small unintended differences in temperature between different growth chambers, or to differences in the quality of the media that cannot be perfectly controlled by the experimenter. Some of the measured variation may not even be real, but due to the equipment that we use to measure growth rates not being 100% precise (see Section 11.2). Hence we can divide variation into variation due to **factors** of interest, and **random variation** or **noise**. Of course, whether variation is regarded as due to a factor of interest or to random variation depends on the particular question being asked. If we are interested in whether different strains of a parasite cause different amounts of harm to hosts, the ages of individual hosts is something that we are not really interested in, but could nevertheless be responsible for some of the variation we observe in the experiment. However, if we are interested in whether parasites cause more harm in older hosts than in younger hosts, the variation due to age becomes the focus of our investigation.

In very general terms, we can think of life scientists as trying to understand the variation that they see around them: how much of the variation between individuals is due to things that we can explain and are interested in, and how much is due to other sources. Whenever we carry out an experiment or observational study, we are either interested in measuring random variation, or (more often) trying to find ways to remove or reduce the effects of random variation, so that the effects that we care about can be seen more clearly. Life scientists have amassed a large array of tools to allow us to do this, and whilst the multitude of these and their specialist names can appear daunting, the ideas underlying them are very simple. In this chapter, we outline the general principles of experimental design, replication, and sampling, and explain how these can be used to remove the masking effects of random variation.

> Generally, we want to remove or control variation between subjects due to factors that we are not interested in, to help us see the effects of those factors that do interest us.

4.2 Replication

Random variation makes it difficult to draw general conclusions from single observations. Suppose that we are interested in whether male and female humans differ in height. That is, we want to test the hypothesis:

Sex has an effect on human height.

An admittedly left-field way to test our hypothesis might be to find the heights of Pierre and Marie Curie from historical records, and find that Marie was shorter than Pierre. Does this allow us to conclude that human females are shorter than males? Of course not! It is true that some of the difference in height between Pierre and Marie will be due to any general effects of sex on human height. However, there will be a number of other factors that affect their height that are nothing to do with sex. Maybe Pierre had a better diet as a child, or came from a family of very tall people. There are numerous things that could have affected the heights of these two people that are nothing to do with the fact that Pierre was male and Marie female. The difference in height that we observe between Pierre and Marie might be due to the factor that we are interested in (sex), but could equally just be due to other factors.

The solution to this problem is to sample more individuals so that we have replicate males and females. Suppose we now go and measure ten married couples and find that in each case the man is taller than the woman; our confidence that men really are taller

Replication involves sampling and taking the same measurements on a number of different subjects. Indeed, subjects are often called **replicates** and their number the *number of replicates*.

Q 4.1 Are we safe in restricting our sample to married couples?

than women on average would increase. The reason for our increased confidence is not very profound. Common sense tells us that, while it is easily possible that differences in height between a single male and a single female could be due to factors other than sex, it is very unlikely that chance will lead us to select ten out of ten couples where the man was taller, when actually couples where the woman is taller are just as common in the population. This is as likely to occur by chance as tossing a coin ten times and getting ten heads. It is unlikely that in all the chosen couples the males had better diets as children than the females, or all the females came from characteristically short families (unlikely but not impossible, see Section 1.4.2 and Chapter 7 for further consideration of such *confounding factors*). If we found the same pattern in 98 out of 100 couples we would be even more confident. What we have done is **replicate** our observation. If differences were due only to chance, we would not expect the same trend to occur over a large sample, but if they are due to real differences between males and females we would expect to see consistent effects over replicate measurements. The more times we make a consistent observation, the more likely it is that we are observing a real pattern. All statistics are based on replication, and are really just a way of formalizing the idea that the more times we observe a phenomenon the less likely it is to be occurring simply by chance. See Box 4.1 for a more detailed treatment of the relationship between random variation and replication.

BOX 4.1 A detailed example of variation and replication: does genetic modification to chicken feed make egg shells thinner?

Given the importance of variation and replication it is worth spending some time thinking about these issues with the use of an example.

Imagine someone working for the research arm of an animal feed company that is considering whether to add grain that has been genetically modified (GM) to their standard chicken feed. However, it has been suggested that the particular type of genetic modification being used may have a negative effect on the shell quality of eggs laid by chickens, making them thinner and more fragile. How should they decide whether this is the case or not? Hopefully, given that you are reading this book, the answer is obvious to you: they should do an experiment.

The researcher will immediately be faced with the problem of biological variation and how to deal with it. Egg shells, like almost everything we might measure in biology, vary in thickness from egg to egg. The thickness of a shell will vary for many reasons. One of these reasons might be whether or not the hen had been fed GM feed, but the thickness of a particular egg shell may also depend on the time that it is laid, the age of the hen, the breed of the hen, how much water she has drunk, the temperature of the hen house ... the list goes on and on. The researcher's job is to try to understand this variation and determine how much, if any, of it is due to the feed the hen has been given. Or to put it another way, the research needs to test the null hypothesis:

The thickness of an egg shell is not influenced by whether the hen that laid it has been fed standard or GM feed.

Let's start by thinking about hens that have been fed the standard feed. Imagine we have five such hens. We take one egg from each hen (why only one? ... this is discussed in detail in Chapter 5) and measure the thickness of the shell. We can be extremely confident of one thing, the thicknesses of these egg shells are very unlikely to be the same—egg shell thickness, like almost everything in biology, varies. We can do our best to reduce the variation by standardizing as many other factors as possible. So we might choose to use hens of the same breed, of the same age, and kept in identical conditions. We might choose eggs laid on the same day, and we would certainly try to make sure that we measured all eggs using the same procedure. Reducing variation in this way is one of the reasons that we carry out experiments in carefully controlled conditions. However, no matter how hard we try to standardize everything, it is still likely that the shells will vary to some degree. Good experimental technique can reduce variation, but it will rarely completely remove it.

Now let's get a picture of what this variation looks like. Imagine that we had the shell thickness measures for literally hundreds of thousands of eggs from hens fed standard feed. If we looked at the distribution of this data the chances are that it would look a lot like one of the distributions in Figure 4.1.

We would likely find that the measures clustered around a mean value, in this case 3 mm, but many are slightly smaller or larger than 3 mm and a few are much smaller or much larger. Statisticians call this kind of bell-shaped curve a *normal* or *Gaussian* distribution, and lots and lots of biological variation looks like this. As well as a mean, every Gaussian distribution has a **standard deviation** (SD), which is a measure of how spread out the data is around the mean. In Figure 4.1 you can see that both distributions cluster around the same mean, but the top distribution is more spread out around the mean value (it has a higher SD, or is more variable). The smaller the SD, the more tightly the data are clustered around the mean and the less likely we are to see extremely high or low values. If you need convincing of this, look at the two distributions again. We can ask in each case, if we pick an egg at random, how likely is it that it will be thicker than 4 mm or thinner than 2 mm (that is, more than 1 mm different from the mean)? To answer this question we need to know what percentages of egg shells in the two distributions are greater than 4 mm or less than 2 mm. We have shaded the sections of the two distributions corresponding to these egg shells. You can see that the shaded area on the top distribution is larger (and so corresponds to a larger fraction of egg shells) than those on the bottom distribution, so the more variation among our egg shells, the more likely we are to pick an unusually thick or unusually thin egg shell in our sample by chance.

We could then do the same for eggs from hens on the GM diet. The pattern of variation will probably look much the same, indeed if feed type actually has no effect on shell thickness it will look exactly the same. However, if feed type really does have an effect on shell thickness, our distribution would be clustered around a different mean. Figure 4.2 illustrates these two scenarios diagrammatically.

In the top panel the null hypothesis is true and GM grain has no effect on shell thickness. This means that the underlying distribution of shell thickness from hens

The **standard deviation** (SD) is a measure of the spread of values around the mean (or average) value. If the distribution of values is a Gaussian shape, then approximately 95% of the values are within two SDs of the mean. So if the mean thickness of shells is 3 mm and the SD is 0.5 mm, then roughly 95% of shells will have a thickness between 2 mm and 4 mm.

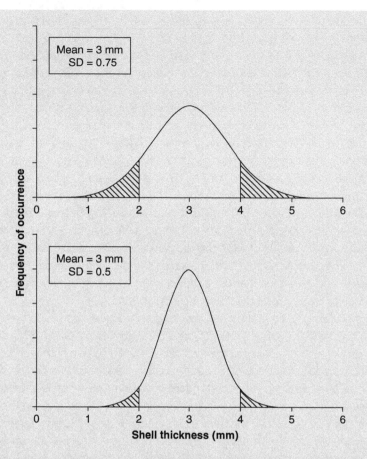

Figure 4.1 Two Gaussian or normal distributions to illustrate the kind of variation we are likely to see if we were to measure the shell thickness of many hundreds of thousands of eggs from hens fed a standard diet. Both distributions cluster around the same mean value (3 mm), but the distribution in the upper panel has a larger standard deviation, because the variation of the values around the mean value is greater. In the upper distribution, approximately 18% of the eggs have shells that are thicker than 4 mm or thinner than 2 mm, whilst in the lower distribution this is only 5%.

fed the two different diets is exactly the same. In the lower panel the null hypothesis is false and the GM grain does reduce the average thickness of egg shells by 0.2 mm. The distribution of egg shells for the hens on the GM and standard diets are still both Gaussian. The distributions also have the same SDs, as the type of food does not affect the underlying variation of shell thickness within experimental groups. However, the two distributions now have different means. We can think of these distributions as representing populations. One population we might call 'All the egg shells of hens fed a standard diet', or 'standard eggs' for short, whilst the other we might call 'All the egg shells of hens fed a GM diet', or 'GM eggs'. If the researcher could measure every shell in these populations then they could simply say whether their means differed without resorting to any statistics. However, in general, and for obvious reasons, it is not usually possible to carry out that kind of study. Instead we carry out experiments where we examine relatively small samples drawn from the populations that we really want to know about and use the information from these samples to try to say something about the populations that they come from. In this case we would measure samples of eggs from hens fed either standard feed or GM feed and use these samples to estimate the means of the populations they are drawn from. We can then use an appropriate statistical test to examine whether the difference between the means of our samples suggests that the populations that they represent have the same mean as each other or different means.

To illustrate the importance of replication in this process, we are now going to carry out some imaginary experiments using a computer and a random number generator. We will start by assuming that we live in a world like the lower panel in Figure 4.2 and the null hypothesis is actually false, because GM feed does affect

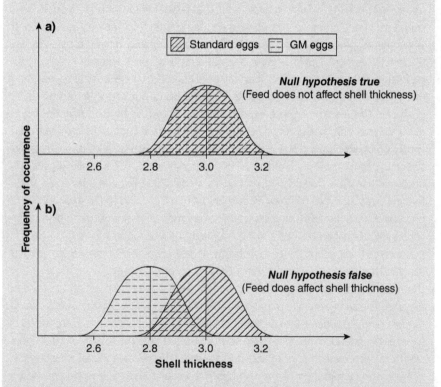

Figure 4.2 The distributions of egg shell thicknesses under two different scenarios. Under both scenarios shell thicknesses follow Gaussian distributions but in the top panel, the two populations have the same mean because the treatment is having no effect on shell thickness (i.e. the null hypothesis is true), whilst in the bottom panel the treatment is affecting shell thickness (i.e. the null hypothesis is false) and the means of the distributions differ.

the thickness of shells, reducing them by 0.2 mm on average. Now let's assume you have ignored all our advice on replication and compare only a single standard egg to a single GM egg. To recreate this experiment on our computer we will draw a single random number from a normal distribution with a mean of 3 mm and a SD of 0.2 mm. This will be the measure for our standard egg. We will then repeat the process to obtain a measure for our GM egg, but this time drawing from a normal distribution with the same SD but a mean of 2.8 mm. The data from our experiment is standard egg = 3.17 mm and GM egg = 2.85 mm, suggesting that the GM feed is reducing egg shell thickness by 0.32 mm. However, now let's imagine our colleague does the same experiment in the same way (computer, two more numbers please!), and their data is standard egg = 2.80 mm, and GM egg = 2.77 mm. These values suggest an effect of GM of −0.03 mm, much smaller than the value that we got in our experiment. Because we know what the populations really look like, we can see that this is because our colleague has measured a relatively small standard egg. However, when we are doing experiments for real we do not have that luxury of knowing the underlying population distribution (indeed, that is what we want to deduce from our experiment) and so are left wondering who is right (in fact, we can see that both 'experiments' produce an estimate that is quite a long way from the true value of 0.2 mm difference). Now the great thing about doing these kinds of imaginary experiments on computers, rather than with real hens, is that it is easy to repeat them many, many times, So let's now create 10 000 runs of our 'one egg from each' experiment. The results of these experiments are shown in the top panel of Figure 4.3. The first thing that should be clear from this figure is that, although the experiments do cluster around the true difference of −0.2, there is a lot of variation between experiments. In fact, of the 10 000 experiments, 2427 actually suggest GM feed increases the thickness of egg shells, whilst 2351 experiments suggest GM feed reduces egg shell width by more than 0.4 mm (i.e. twice the real value). So the first obvious problem with an unreplicated experiment is that it will generally provide very unreliable estimates of the things we might be interested in.

Now let's think about things in a slightly different way. This time we are going to assume that we are in a world where GM has no effect on egg shell thickness, so the null hypothesis is true. Putting this more formally, the population of standard eggs has the same mean (3 mm) and SD (0.2 mm) as that of GM eggs. Let's use our random number generator to carry out some imaginary experiments in this world by again drawing random numbers from normal distributions. The results of 10 000 'one egg from each' experiments in this world are shown in the lower panel of Figure 4.3.

Again, this histogram is centred on the true value (in this case zero, because the populations have the same mean), but is highly variable, with some of our experiments suggesting very large effects of GM (both positive and negative), when we know there is really none (remember, our samples have come from distributions with the same mean). To put this in context, the result that we got in our first experiment when the null hypothesis was false suggested that the effect of GM might be as big as −0.32. However, looking at Figure 4.3 it is very clear that many of our imaginary

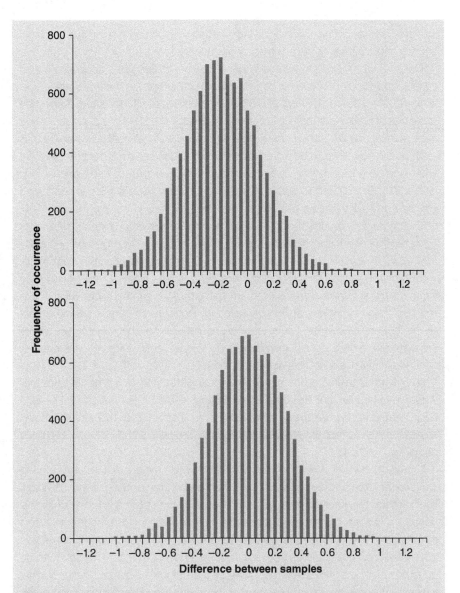

Figure 4.3 The results of 10 000 imaginary 'one egg from each' experiments in a world where GM has no effect on shell thickness (bottom panel) and another world where GM reduces shell thickness by 0.2 mm (top panel). The height of each bar represents the number of experiments that estimated a particular difference (rounded to the nearest 0.05 mm). In both worlds, the estimates from the experiments cluster around the true difference between the population means in that world, but in both cases there is considerable variation among experiments.

experiments generated differences as big as −0.32, even when there really was no difference between the two populations. In fact, 1274 of our 10 000 imaginary experiments produced a difference as big as this. With small samples (and our sample size of one could not be any smaller), large differences between samples experiencing different treatments will often occur by chance alone.

The amount of variation we see in the experiments described depends on how variable the populations we are sampling are. For example, in the experiments described, SD = 0.2, and an effect of 0.32 was quite likely to be measured, even when there was actually no difference in means. However, if the SD of the populations had been smaller, say 0.05, then observing a difference as large as 0.32 when there is really no difference would be extremely unlikely. You can see this effect in Figure 4.4 where we have repeated the previous thought experiment but using populations with a SD of 0.05. Of our 10 000 imaginary experiments, not a single one suggested GM reduced shell thickness by 0.32 mm or more.

Thus, if we have an idea about the underlying variation, we can actually say something sensible about our chances of seeing different experimental results under different hypotheses. However, when we do an experiment we generally don't know the SD of the populations we are drawing our samples from. Nevertheless, if we sample several individuals in each treatment group we can estimate variability in the population by looking at how variable individuals in our sample are (in simple terms: if all are very similar, it suggests the population is not that variable; but if they are quite different, the population is probably very variable). With only one egg per treatment we can say nothing about this variation.

So how does taking replicate measures deal with these problems? Let's repeat what we did earlier, but this time we will measure the eggs from several hens in each treatment and use the difference between the means of these two samples to estimate the effect of GM. Let's look at what happens with 5, 25, and 100 hens per group (Figure 4.5).

The first thing that should be apparent is that, while there is still variation among experiments, there is less variation than when we only measured a single individual. Furthermore, the variation gets less as the sample size gets bigger. Why does this happen? The reason is easy to understand. If we draw a single number from a normal distribution there is a reasonable chance it will be unusually high (or unusually low). If we draw five numbers from a normal distribution, then for their mean to be unusually high, it is necessary that most of the numbers are also unusually high, which is less likely to occur. If we draw 25 numbers, then the chance of the mean being unusually high becomes even less. Another way to think of this is that when we draw several numbers and calculate a mean, whilst some of the numbers will be likely to be unusually high, these will be cancelled out by other numbers being unusually low. In practical terms this implies that our estimates of the population means become more reliable as our samples get bigger, and the difference between sample means becomes a more reliable indication of any difference between population means.

Similarly, in a world where GM has no effect, we have already seen that experiments based on single samples will often show large differences, simply due to chance. As the sample size increases the variation decreases, and the chance that

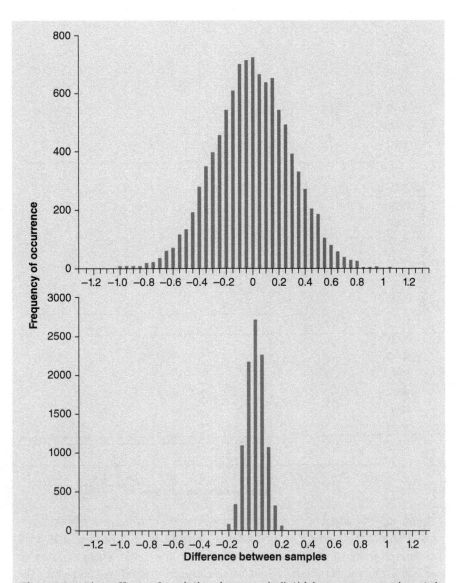

Figure 4.4 The effect of variation in egg shell thickness on experimental estimates. Estimates of the effect of feed type on shell thickness from 10 000 single individual experiments. In each experiment, the effect of GM is estimated by subtracting the thickness of the standard egg shell from the thickness of the GM egg shell. The height of each bar shows the number of experiments that gave a particular effect size (rounded to the nearest 0.05 mm). In both cases there is no effect of GM, and the distributions are centred on zero, but in the top panel the underlying variation in egg shell thickness is greater than in the bottom panel.

we will see a large difference between the means of our treatments if GM really has no effect gets smaller and smaller.

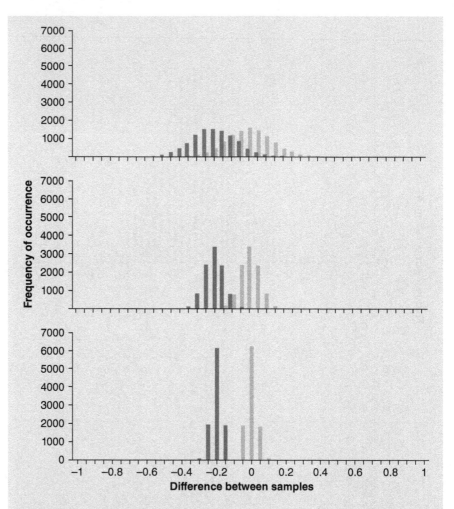

Figure 4.5 The effect of replication on the estimate of treatment effects. Each panel shows the distribution of treatment effects (to the nearest 0.05 mm) estimated from 10 000 imaginary experiments in a world where GM has no effect on eggs shell thickness (light bars), and in an alternative world where GM reduces shell thickness by an average of 0.2 mm (dark bars). In each experiment the effect of GM is estimated by subtracting the mean thickness of the standard egg shell in the sample from the mean thickness of the GM egg shell in the other sample. The sample size of the experiments increases from top to bottom, with five hens per group in the top panel, 25 in the middle, and 100 in the bottom. As sample size increases, the estimate of the treatment effects becomes more precise, and we see fewer experiments yielding extreme estimates. In practice, this means that it becomes easier for us to discriminate between these two worlds from the result of a single experiment.

So replication has two important effects on our experiment; by taking several measures and then taking a mean, some of the random variation is effectively cancelled out making any real differences easier to see. At the same time, experiments based on several replicates make extreme experimental outcomes less likely to occur by chance, and so if we do see a large difference between our treatments, we can be more confident it is because the population means really do differ. Finally, having replicate measures allows us to estimate the variation in the underlying populations too.

→ Replication is a way of dealing with the between-individual variation due to the random variation that will be present in any life science experiment. The more replicates we have, the greater the confidence we have that any difference we see between groups is due (at least in part) to the factors that we are interested in and not due to chance alone.

4.3 Selecting a sample

As discussed in Chapter 1, in general we cannot collect measurements on all the individuals in the population of interest. Rather, we focus our study on a sample that we hope will be representative of the population that we are interested in. So if our question was to estimate the height difference between ten-year old Scottish boys and girls, it would be impractical to measure the heights of all approximately 75 000 10-year-old children in Scotland. We will leave to Chapter 6 the question of how many boys and girls would be appropriate, but you might imagine samples of say 100 of each would be reasonable. What we are interested in here, having decided on an appropriate sample size, is how we select individuals from the population to be in the sample. Notice that in some experiments, having drawn a sample of individuals, we must then allocate them to different groups that experience different treatments within our experiment; we will leave discussion of such allocation strategies to Chapter 7, and focus here on drawing the initial sample. We will describe a number of the most commonly encountered methods.

@ Some further methods (specifically snowball sampling, systematic random sampling, and proportional quota sampling) can be found in Section 4.3 of the supplementary information. Go to: **www.oxfordtextbooks.co.uk/orc/ruxton4e/**.

4.3.1 Simple random sampling

The simplest method of selecting a sample is called *simple random sampling*. Imagine that from some government census data we obtain a list of all the ten-year-old girls in Scotland: 37 467 of them. We simply put them in a spreadsheet and assign a number from 1 to 37 467 to each. To obtain a simple random sample of 100 of these we then ask the computer to select a random number from 1 to 37 467 where each number has an equal probability of being selected. Let's say that number is 21 456; the girl with that number then becomes the first person in our sample. We repeat this process until we have 100 unique numbers (i.e. we have no duplicates), and so we have identified 100 girls chosen entirely at random and independently of each other. That is our simple random sample.

If your population is smaller then you can get a random sample without using a computer. Imagine we have a herd of 57 cattle and we want to sample 15 of them. Each cow has a unique identification number embossed on a tag on its ear. We simply write each of these numbers down on 57 identical pieces of card, drop the pieces of card into a bag, stir the cards round with our hand for a bit, then pull out 15. This should give a simple random sample. The key thing for the two techniques described above is there should be no way the Devil's Advocate can be concerned about unconscious bias in which individuals are selected for our sample. You could simply make a list of the 57 cows on a sheet of paper, close your eyes then stick your finger onto the piece of paper and read-off the ID number nearest to your finger. You could repeat this until you had 15 unique numbers. This is marginally less work that the 'card in a bag' approach but it introduces risk of bias. Subconsciously if you stuck your finger high on the list of numbers one time, you are likely to go lower on the list next time. This is not in the spirit of random sampling. The key demand of random sampling is that at any point during the selection of your sample every individual who might be added to your sample next (that is who hasn't been picked already) has an equal probability of being the next one picked. That is true of the first two methods we describe, but in this last method it is not because we think there is a risk that after a number from low on the list has been selected the next number selected is more likely to be a high number through your unconscious effort to spread your pointing around the list. So the secret to good random sampling is to remove human factors from the selection process so there can be no risk that someone accuses you of bias in your sampling.

Notice you can always do simple random sampling as long as you can list all members of the population that you want to sample from. In some cases, such as herd of 57 cows, this requirement is pretty easy to meet. But the example of all ten-year-olds in Scotland is an example of a case where this will be less straightforward. Census data of the ages of all residents of Scotland probably exists in some government department; but obtaining access to this data may be difficult or even impossible. If it is possible then this census data will have been collected some time ago and so will not be completely accurate; you have to weigh up how important this inaccuracy is. For example, imagine the census occurred 12 months previously, obviously all the children who were nine then will be ten now; but there will be some children who have died or left the country and some other children who have come to the country only in the last 12 months. However, these unusual cases will likely only lead to a very small inaccuracy in using the 12-month-old census and you can probably defend this approach to a Devil's Advocate.

➡ In a *simple random sample* we list all members of the population of interest, then independently and randomly select a fixed number of them to be in the sample; all have equal probability of being included.

4.3.2 Remember, you want a representative sample

You want a representative sample, and most but not absolutely all random samples are representative. Imagine in the herd of 57 cows there are 20 three-year olds,

19 four-year-olds, and 18 five-year-olds. Imagine now that by chance we notice that our random sample of 15 individuals contains 14 three-year-olds and 1 five-year-old. This seems a very uneven distribution across the age range—is this a problem? Possibly. If we think that the traits that we want to measure on our sample of cows are likely to be influenced by age then the fact that our sample age structure is very different from the age structure of our population of interest is a concern. In this case we should simply reject this sample and draw a fresh sample of 15 that is very unlikely to have such an unusual age distribution. However, such situations will very, very infrequently occur; almost all random samples will be representative. The larger the sample the less likely it is that a random sample will be unrepresentative. You might go through your whole scientific career and never need to reject a random sample on the grounds of it being unrepresentative.

However, eyeball your sample to check that it is representative before you actually carry out measurements on the sampled individuals. One reason to do this early in the process is to save the time, effort, expense, and animal suffering of taking measurements that you will throw away. The other reason is to avoid accusations that you are resampling because the first sample was inconvenient rather than unrepresentative. Imagine that you measured the traits of interest on the 15 cows in your sample, then carried out statistics on your data and got an answer different from the one you were expecting; you then noticed this issue of unusual age distribution and decided to repeat your experiment with another sample. The Devil's Advocate will surely say that you would not have resampled if you had gotten the result you wanted the first time around, so you are repeatedly sampling until you get the result you want. You can see that this is very poor scientific practice—in fact some people would call it scientific misconduct. One way to avoid an unrepresentative sample would be to use stratified sampling, which we'll discuss next.

➡ You aim to have a sample that is representative of the population. Almost all random samples are, especially for larger samples, but very, very occasionally you can get an unrepresentative random sample. Throw it away and pick a new one.

4.3.3 Stratified sampling

Let us return to the example of the herd of 57 cows. Imagine that we had considered before sampling that the traits we are interested in are likely to be strongly age-dependent, and so we very much want the age structure of our sample to reflect that of the statistical population (the herd). We can do that by stratified sampling. We note that the cows fall into three age classes and that each of these has approximately one third of the cows. We then divide our population up into these three strata (three age classes) and use simple random sampling on each of these three strata independently. That is, to get our sample of 15 we draw a sample of five by simple random sampling from the group of 20 three-year-old cows, five from the group of 19 four-year-olds, and five from the group of 18 five-year-olds. This is a little more work to organize but ensures that, for the trait that we wanted to stratify on (age), the distribution across the sample will be similar to the distribution in the underlying population.

You can stratify on any trait you can measure on subjects, not just age. Consider the case of sex differences in ten-year-olds in Scotland. Here we might have considered that ethnicity might be important to height. The same census data discussed above might also give us the stated ethnicity of each ten-year-old. We could then decide to stratify on that ethnicity. Imagine the distribution of different ethnicities in ten-year-old children was 96.0% White, 2.7% Asian, 0.7% Black, 0.3% Mixed, 0.2% Arab and 0.1% Other. If we were sampling 1000 girls we could split the girls into groups (strata) by ethnicity and for a sample of 1000 select 960 from the 'White' group, 27 from the 'Asian' group etc. This would ensure that the distribution of ethnicities in our sample is similar to that of the underlying population.

In the case above where ethnicity will not have a huge effect on height (Black ten-year-olds will not be twice as tall on average as Asian ones) and where there is one dominant stratum (96% are White) then there is not a strong need to stratify. However, if you stratify correctly then there is no drawback to stratified sampling except for the extra effort involved. Where stratification might be very valuable is if a small fraction of the population might be very influential. For example, imagine that we were sampling positions in a large forest where we will place pitfall traps in order to measure terrestrial invertebrate biodiversity (see Figure 4.6). The margins of the forest might be a relatively small fraction of the overall land cover of the forest but might be very high in biodiversity compared to the interior. By stratifying we can make sure that our random sample includes an appropriate (small) fraction of these sites that we think might be uncommon but influential. If we selected sites by simple random sampling we might not select any sites near the margins and so might not have the ability to evaluate this part of the study population that we think might be important.

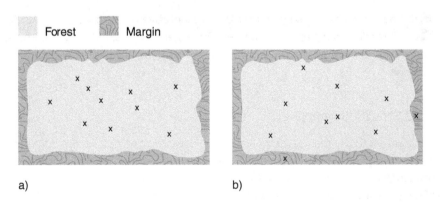

Figure 4.6 Comparison of random and stratified sampling schemes. If we place traps at random as in a) then it is quite likely that none of our 10 traps end up in the 20% of the forest that we classify as margin habitat. With stratified sampling, we allocate 2 of our 10 traps randomly in this marginal area and the remaining 8 in the interior as in b). This maximizes the chance that our traps provide a representative sample of the invertebrates in the whole woodland.

It is possible to stratify on more than one variable. In the 'Scottish children' example we could stratify on geographical area as well as ethnicity, so that of the 960 White girls we ensure that (for example) 8% of those come from the city of Edinburgh.

Stratified sampling requires that you have good information on the distribution of the variable that you want to stratify on. There is no danger in stratifying carefully on a variable that turns out not to be influential—if ethnicity turns out to not be a good predictor of the height of ten-year-olds then our sample is not any less valid because we chose to sample by stratifying on that variable. But, if the variable is influential and we stratify inappropriately then we can end up with a biased sample. Say that marginal sites in the forest in fact represent 7% of the land area, but we mistakenly thought that they made up 17% of the land area: if those sites end up making up 17% of our sample sites then our sample is not a good representation of the forest, because too much weight is given to marginal sites. Thus, you should only think of stratified sampling when (i) you have identified an influential variable that you would like to make sure is appropriately represented in the sample; and (ii) you have good information on the distribution of that variable across the statistical population. If not, then stick to simple random sampling.

➜ We use *stratified sampling* to ensure that the sample accurately reflects the population for a trait (or traits) that might be highly influential on the qualities we want to measure on subjects in our sample.

4.3.4 Cluster sampling

Imagine that we had used simple random sampling or stratified sampling to identify 1000 girls and 1000 boys in Scotland to be in our sample. These individuals will be scattered right across the country. Collecting their heights will take a lot of organization and a lot of driving. With a bit of organization by phone we can arrange to measure geographically close children on the same day. Having measured one child, we get into the car and program our satellite navigation device to help us drive to the next individual; we can then sit in the car and wait with our flask of coffee until the time we have arranged to measure them. Overall, we will spend a lot more time driving and drinking coffee than we will measuring children. This may be worth the effort, but sometimes we can still get a good sample and cut down on organization and logistics by sampling not single individuals but natural clusters. This is the essence of cluster sampling. It doesn't have any theoretical advantages over simple random sampling, but it can sometimes give you just as good a sample at much reduced effort. But it requires that individuals fall into natural clusters.

Ten-year-old Scottish children do fall into natural clusters. They can naturally be grouped into schools. On average a primary school might have 30 ten-year-old children so rather than randomly selecting 2000 individual children from across Scotland in a simple random sample, in cluster sampling we might use simple random sampling to select 70 primary schools and measure all the children in each school. This will greatly reduce the hassle of data collection: we will have to do a lot less driving. Is this still a representative sample of Scottish ten-year-olds? Probably. The average height of girls in a school will be affected by things like the ethnic and socioeconomic breakdown of the local area, so we need to be sure that we average across this variation; by sampling 70 schools we might reasonably expect to do this. But we must also be careful that when we adopt cluster sampling we do not subtly but significantly change the population from which we sample. By sampling at school level we will no longer include children who are taught at home rather than at school in our sample. In fact only 800 children (of all ages) in Scotland are taught outside the school system, so this change in sampled population should not concern us much. But, as a generality, we need to be careful of changing the underlying population (and so changing the specific question we are investigating) when we adopt cluster sampling for reasons of convenience.

It is possible to imagine hybrid sampling schemes that involve layers of sampling. For example, if we decided that 2000 children felt like a good number to aim for but 70 schools was a little small then we could select 100 schools then within each school use simple random sampling to select ten ten-year-old boys and ten ten-year-old girls. Notice that very small schools might have fewer boys and girls in total than you need for your sample, and you might decide either just to sample all of them (and visit more schools); or to avoid such schools. The best choice will depend on trading off practicality with any concern of introducing bias by excluding some types of school. This question of whether to sample more schools or more pupils within each school is another manifestation of the concept of subsampling discussed in Section 11.2.1.

Imagine that you wanted to ask a different research question that required to you to allocate children in your survey to different treatments. Say you might be interested in investigating which of two methods was most effective at teaching children a foreign language; or you might be interested in which of two toothpastes was more effective at preventing tooth decay in children. You could either allocate whole clusters (whole schools) to a treatment, or allocate individuals to a treatment. This decision will come down to practicality and risk of interaction between individuals. For the teaching method it would be much, much easier to implement if allocation was done at a whole-school level and all of the children in one school were allocated to the same method. This would also help with avoiding concerns stemming from interaction between individuals. Since the children will interact with each other outside the classroom, sharing knowledge, we can see that if half the children in the school were taught one method and half the other, then the observed difference between children on the two schemes might well be reduced because when the children interact with each other the children experiencing the better teaching method help raise the performance of those on the other method. However, for the toothpaste study we think the chance of children sharing toothpaste is pretty low and so each child can be more readily considered as independent, and there is no great organizational challenge associated with assigning individual children in the same school to different toothpaste types (particularly if clever design of packaging (see Section 2.3.1 on placebos) reduced obvious differences in appearance between the two types).

Lastly, we recommended that clusters (schools) be selected by simple random sampling where each school has an equal probability of being in the sample. You will also see cluster sampling implemented where the probability of each cluster being in the sample is weighted by the size of the cluster. In our case that would strictly be by the number of ten-year-olds in the school, but the total number of children at the school is probably easier to obtain information on and would do just as well. Weighted sampling is a little trickier to implement but by weighting by cluster size you can see that we get back to the underlying assumption of simple random sampling of individuals, that all individuals have an equal probability of being in the sample. We mention this idea of weighting only because you might come across it in published papers. For your own work we do not see a compelling reason to add this extra mathematical complexity (and thus extra room for a mistake) to your method of sampling.

 Cluster sampling can be a good way to get a good sample at reduced effort when individual subjects are naturally grouped into clusters.

4.3.5 Convenience sampling

Especially early in your scientific career, when you don't have a lot of time and money to devote to projects, you will often find yourself doing another type of sampling called *convenience sampling* (also called *accidental*, *grab*, or *opportunity sampling*). Bluntly, this sort of sampling is not as methodologically pure as those already discussed, but (as the name suggests) it is often just an awful lot more convenient. It might not be as appealing to the purist as the methods discussed already, but it can still be done really well (or really badly!) You would like to estimate the difference between the heights of ten-year-old Scottish boys and girls, but you simply don't have the resources to travel across the country and conduct sampling in any of the ways discussed previously. This doesn't mean that you cannot carry out a useful study. It occurs to you that there is a local zoo that attracts families and school groups from across Scotland (and beyond), and you approach this zoo and obtain permission to try and recruit children into your study. At the zoo you station yourself somewhere where visitors pass by, and when a group with at least one child that you think might be ten years old passes by you approach an adult in the group (likely a parent, school teacher, or youth group leader) and explain the nature of your study and whether the child might be appropriate for your study (correct nationality and age). If they are suitable and all appropriate parties are willing then you add this child to your sample. In doing this you can quite quickly build up a substantial sample that will let you explore the question of interest to you.

However, two issues you must be aware of are *selection bias* and *external validity*. It can be easy to introduce selection bias in convenience sampling and you must guard against even the appearance of this. Selection bias occurs when some suitable subjects are more likely to end up in your sample than others in a way that biases your results. Set yourself careful protocols for selection that will help you avoid selection bias and explain to others that you have avoided bias. In this case, the issue is that you are going to approach groups that might have a child who is ten with a view to recruiting that child. The real danger for this study is that your assessment of whether a child looks like they might be ten could in part be based on their height, the very factor you are interested in measuring. It would be very unfortunate if you let suitable subjects pass by unrecruited because you guessed they were younger than ten because they are unusually short for a ten-year-old. The way around this would be to approach any group where it is even slightly possible that a child might be ten, so you steel yourself to approach any group whether at least one child looks as if they are definitely older than five but less than fifteen. This will cause you extra effort and occasionally some embarrassment, but will help you to avoid bias.

We should mention another type of bias related to humans: *recruitment bias*. Until now we have assumed that all individuals approached to be in your study agree to be in the study, or equivalently that there is no relation between the trait we are interested in (height) and willingness to be in the study. If we break this assumption then we have recruitment bias. Here we have to be careful; you could imagine that a child who is self-conscious about their height because they perceive it to be unusual might be less willing to take part in the study. Again, it is worth thinking through a protocol that can help avoid recruitment bias and demonstrate to others that you have avoided recruitment bias. In this case, it would be worth emphasizing to the potential subject that (i) they were approached because they looked like they might be ten and not because there was anything unusual about their height; (ii) that you won't record their name; (iii) that you will measure their height in a way that means other people will not see the value; and (iv) you will never disclose their value for height to anyone. There is no guarantee that such steps will entirely eliminate recruitment bias but they should help reduce this concern considerably.

External validity is the extent to which the study can be generalized to the population of interest. The previous methods have very strong external validity because all members of the population could potentially be in the sample. In our child-heights example, all Scottish schoolchildren might have been included in the sample; when we use a convenience sample at only one location you have to make more of an effort to make people believe that your study has external validity. In the child-height study it might be worth asking participants about their ethnicity and about where in Scotland they come from. If you can show that your sample does actually have a wide spread geographically, and has an ethnic distribution that is similar to the population of interest (all Scottish ten-year-olds), then this should help convince others about the external validity of the study. External validity is another way of thinking about how appropriately you can generalize from a study—see Statistics Box 3.1 for more on this. (It won't surprise you to find that there is also a piece of experimental design jargon called *internal validity*. This is about the validity of any inferences you make about cause-and-effect relationships, and so is about avoiding confounding factors—see Sections 1.4.2 and 3.1.3.)

 Convenience samples are much more convenient to obtain, and can still be useful if you take care in the design of your data collection.

4.3.6 Self-selection

Self-selection is a real limitation to the usefulness of the phone polls beloved of news-papers and TV stations. Generally, people are encouraged to phone one number if they, for example, prefer Coke to Pepsi or another number if they prefer Pepsi. Now, what can

Particularly in studies involving humans, there is a blizzard of terminology associated with biased sampling: selection bias, volunteer bias, non-response bias, ascertainment bias, detection bias, information bias, response bias, assessment bias, observer bias, recall bias, allocation bias, and attrition bias. See Section 4.3.5 of the supplementary information for brief definitions if ever you encounter any of these terms in other reading. Go to: **www.oxfordtextbooks.co.uk/ orc/ruxton4e/**.

you conclude if the Pepsi number receives twice as many calls as the Coke number? First you must think about the population that is sampled. The population that sees the newspaper article with the phone number is not a random cross-section of the population of people who drink cola. People do not select newspapers in a newsagent at random, so immediately you are dealing with a non-random sample of the population. But things get worse. People who read the newspaper do not all respond by picking up the phone, indeed only a small fraction of readers will do so. Can we be sure that these are a random sample of readers? In general, this would be very risky; generally only people with a very strong opinion will phone. So we can see that these phone polls are very poor indicators of what the wider public believe, even before you ask questions about whether the question was asked fairly or whether they have safeguards against people voting several times. Our advice is to be very careful when interpreting data that anyone collects this way, and, if you are sampling yourself, make sure you are aware of the consequences of this sort of self-selection.

Q 4.2 What do we mean by asking the question 'fairly'? Can you give examples of unfair ways to ask the question?

In human studies a certain amount of self-selection is almost inevitable. If you decide to ask random people in the street whether they prefer Coke or Pepsi, then people have the right to decline to answer, hence they self-select themselves into or out of your study. Your experimental design should seek to minimize the extent of self-selection, and then you must use your judgement to evaluate the extent to which the self-selection that remains is likely to affect your conclusions.

Self-selection is not confined to human studies. For example, if we are seeking to measure the sex ratio of a wild rodent population then it would seem logical to set out live traps and record the sex of trapped individuals before releasing them. However, this will give an estimate of the sex ratio of that subset of the rodent population that are caught in traps, not necessarily the general population. It could be that some sections of the population are less attracted to the bait used in the traps than others or less likely to be willing to enter the confined space of the trap than others. Hence, the estimate of sex ratio that we get is affected by the fact that individuals may to some extent self-select themselves into our study by being more likely to be trapped. If an individual's likelihood of being trapped might be related to its sex then the sex ratio of the trapped sample might be quite different from that of the wider population.

Q 4.3 What can you do to explore how serious self-selection is for your estimate of sex ratio in the rodent population?

Let's now move on to the next chapter, where we examine replication a little more closely so as to warn you of some potential pitfalls.

➡ Sometimes some self-selection is unavoidable, but try to minimize it, then evaluate the consequences of any self-selection that you have left.

■ Summary

- Whenever we carry out an experiment, we are trying to find ways to remove or reduce the effects of random variation, so that the effects that we care about can be seen more clearly.

- Replication involves making the same manipulations and taking the same measurements on a number of different subjects.

- Replication is a way of dealing with the between-individual variation due to the random variation that will be present in any life science situation.
- In the life sciences, we are almost always working with a sample of experimental subjects, drawn from a larger population. To be independent, a measurement made on one individual should not provide any useful information about the measurement of that factor on another individual in the sample.
- In a simple random sample we list all members of the population of interest, then independently and randomly select a fixed number of them to be in the sample—all have equal probability of being included.
- Simple random sampling (especially for larger sample sizes) almost always gives a representative sample. But once or twice in your career you might have to reject a sample as unrepresentative.
- If you want a sample to be highly representative of the population distribution of a given trait then stratified sampling can achieve this.
- If individual experimental subjects naturally cluster then selecting clusters rather than individuals might be a lot more convenient.
- Often we end up convenience sampling. Generalizing from a convenience sample can be a challenge, but you can meet that challenge with careful sample design.

Pseudoreplication

5

The concept of replication, introduced in the last chapter, implies that you measure a sample of independent experimental subjects. This chapter is a warning about pseudoreplication, where your subjects are not independent of each other but you treat them as if they were.

- We begin by discussing important details of what is meant by independence (Section 5.1).
- We then discuss some common sources of non-independence (Section 5.2), before considering how to deal with non-independence within your sample (Section 5.3).
- In some situations non-independence is inevitable, but this need not invalidate a study if this issue is carefully handled, as we discuss in Section 5.4.
- In fact, pseudoreplication can be linked to the concepts of third variables and confounding factors that we have already discussed, and we consider this link in Section 5.5.
- Lastly, we discuss issues of non-independence in studies that involve looking for change over time (Section 5.6).

5.1 Explaining what independence and pseudoreplication are

As we have seen in Chapter 4, replication allows us to deal with the between-individual variation that is due to noise, and so see any biologically interesting patterns more clearly. However, this effect of replication relies on one critical rule: the replicate measures must be *independent* of each other. What this implies is that any measurement is just as likely to have a positive deviation from the mean due to noise as it is to have a negative one. Thus, if we examine a group of independent individuals their deviations

We are interested in asking questions about a certain group of individuals (the **population**). Generally this population is too large for us to study every individual, so we study a representative subset (the **sample**).

Two replicates are **independent** if the value of one does not change your expectation of the value of the other, given that you already know the distribution of potential values. Non-independent replicates are often called **pseudoreplicates**. If you apply a statistical test that assumes independent replicates to pseudoreplicates, then the test can become unreliable; this mistake is called *pseudoreplication*.

will tend to cancel out, and the mean of the **sample** will be close to the mean of the **population**. This will allow us to make accurate inferences on the population on the basis of our sample. However, if we don't design our experiments carefully, this will not always be the case.

To be **independent**, a measurement made on one individual should not provide any useful information about that factor on another individual. For example, imagine that we wanted to test the hypothesis:

Blue tit nestlings raised in nest boxes suffer more from external parasites than those raised in natural cavities.

If we take the four nestlings in a particular nest box and count the number of parasites on each, then this does not give us four independent measures of the parasite load of nest-box chicks. There is highly likely to be a strong correlation between the parasite loads of nest-mates, so if we find one has an unusually high parasite burden then it is likely that all the others in that same nest box will do too. The problem here is that the parasite burden of a particular chick will not simply be affected by the factor we are interested in (whether the nest site is a human-constructed box or natural cavity) but also by noise factors such as whether the site was occupied in the previous year. If the site was occupied in the last year then this will likely increase the parasite burden seen this year. Previous occupancy is not a factor of interest to us; it is a confounding factor contributing to the noise in our measurements. We would like to average across independent replicates to average out this noise. But here if one of the nestlings in a particular nest site experiences a positive deviation in its parasite load because of previous occupancy then automatically so will all its nest-mates (since they necessarily share the same value of previous occupancy), and so we don't get the cancelling effect of replication that independence should bring us. If replicates are not independent, they are often termed **pseudoreplicates**.

Let us examine this more closely. Imagine we measured 42 chicks from ten nest boxes and for each we had a measure of the number of external parasites counted on it. You could imagine a histogram of those 42 numbers. Or if you lack imagination you could look at the histogram in the top panel of Figure 5.1. Now suppose your field assistant catches another chick, and challenges you to guess its parasite-load. In the absence of any other information your best guess would probably be close to the average value in the histogram (where the bulk of the 42 measured values lie). However, before you guess your colleague lets slip that the chick was from nest box no. 7. The lower panel of Figure 5.1 shows the same histogram, but with the other 4 chicks from nest box no. 7 highlighted. Now what parasite load would you guess for the chick? We suspect that you would now change your guess to something much higher, since the other chicks from box no. 7 all have higher than average parasite loads, and your biological intuition tells you that chicks in the same nest box are likely to have similar numbers of parasites. The fact that this additional information causes you to revise your best guess for the parasite load of the new chick tells you that nest-mates are not replicates but pseudoreplicates of parasite load on chicks in next boxes. Remember we said that, to be independent, a measurement made on one individual

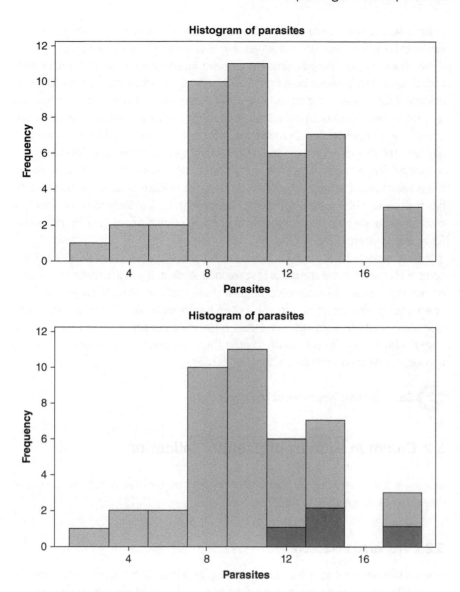

Figure 5.1 Top panel is a histogram of 42 values, each of which is the number of external parasites counted on a blue tit chick sampled from one of ten nest boxes. Bottom panel is identical except that the values relating to four chicks from nest box no. 7 have been identified with darker shading.

should not provide any useful information about that factor on another individual. Our nest-mates fail that test, and so are not independent replicates but pseudoreplicates of parasite load.

Another useful rule of thumb you can use is to ask yourself if two individuals in a sample are expected to be more similar to each other (for the measure of interest) than a random pair of individuals in the sample. If this is true, then we do not have independent samples. Here again, we see that nestlings from the same box are more similar to each other that two individuals drawn at random from the histogram.

We should strive wherever possible to have independent replicates rather than pseudoreplicates, because independent subjects give us more information. Four blue tits from four different nest boxes clearly give us more insight into what nest boxes in general are like, than four blue tits that all share the same nest box. Also, as discussed above, non-independence can make the effect of confounding factors more of a problem. The four chicks in the same nest box will experience the same conditions due to a whole range of possible confounding factors (e.g. whether the box was used in the previous year, the material used in the nest, the amount of sunshine that directly hits the box, and how waterproof the box is); this is much less likely to be true for four chicks in separate boxes, thus the effects of these potential confounders are weaker for independent replicates.

Most simple statistical procedures assume that the replicates in the samples that you compare between in the statistical test are independent. If you use these procedures on non-independent pseudoreplicates then these test can often become unreliable. This inappropriate use of statistics is called *pseudoreplication*. There are statistical procedures that can cope with non-independent data but these are not easy for the beginner to use. Anyway, prevention is better than cure, and so we recommend avoiding non-independence of replicates whenever you can.

 Strive to have independent replicates.

5.2 Common sources of pseudoreplication

Biology is full of sources of non-independence that can catch out the unwitting researcher. Here are a few to watch out for.

5.2.1 A shared enclosure

Imagine that we want to know whether adding enrichment to primate enclosures decreases the level of aggression by the inhabitants. We go to our local zoo where they have two enclosures, each containing ten chimps. We add a selection of toys and playthings to one enclosure in an attempt to enrich the enclosure. We leave the other enclosure unmodified. If we then observed the two enclosures and find that the level of aggression of the ten chimps in the enriched enclosure is significantly lower than of the ten in the other enclosure, is this good evidence that enrichment reduces aggression? Unfortunately it is not, because the behaviours of the ten chimps in a single shared enclosure are not ten independent measures of the effect of enrichment on captive chimpanzees. The problem is that there will be many other differences between the two enclosures: maybe one is warmer or sunnier than the other, or slightly bigger than the other. Thus, chimps in an enclosure share many other factors that are nothing to do with the enrichment and any of these factors could explain the difference in behaviour between the groups.

This type of problem can crop up in any number of situations. This was the issue in the blue tit example discussed in Section 5.1. If your experiment looks at the effect of different temperatures on the courtship behaviour of crickets, but all the crickets for one temperature treatment are in one incubator and all the crickets for the other are in another, then the crickets are not independent measures of the effect of temperature because temperature is confounded by all the other differences between the two incubators. Similarly, in the biomedical sciences it is not uncommon for several experimental rodents to be kept in a single cage rather than being housed individually. If the experimental procedure involves all the individuals within a cage receiving the same treatment (for example, a drug which is administered through their shared water bottle), rodents in the same cage are not independent measures of the treatment since they all share the same cage. A similar effect can even occur in molecular biology when the experimental set up requires that assays are run in physical blocks. Thus suppose you are interested in whether a treatment affects the immune response, and you are going to measure this response using a molecular assay of blood plasma from treated and control individuals. Each assay is carried out in an individual well of a 96-well assay plate, and the entire plate is then placed into a plate reader that quantifies the response in all 96 wells individually. If your study involves you assaying more than 96 samples, so that they cannot all be fitted onto a single plate, and you decide to place your control samples on one plate and treated samples onto another then you are now confounding treatment differences with differences between the assay plates. Perhaps the incubators, cages, or 96-well plates look identical to you; after all they came from the same supplier and have been treated in the same way. Such arguments will not satisfy the Devil's Advocate, who will find all manner of small differences between the most similar-looking cages. And even if a hypothetical supply of identical 96-well assay plates did exist, the Devil's Advocate will simply point out that all samples on a plate are necessarily assayed at the same time and sit in the same space during the period of incubation and so are confounded in time and space.

5.2.2 The common environment

A similar source of pseudoreplication is a common environment. Suppose that you are interested in whether red deer that feed in woodland have higher parasitic worm burdens than those that feed on moorland. You travel to the Highlands of Scotland, find a wood and a moor where red deer are feeding, collect faecal samples from 20 deer at both sites, and determine the number of nematode eggs in the samples. Does this allow you to see whether woodland deer have higher worm burdens than forest deer? While the most obvious difference between the two sampling sites will be the vegetation, there will be many other differences too (perhaps the wooded site is at lower altitude), and any of these differences might affect worm burdens. Thus the deer from each site share many other factors that are nothing to do with the presence or absence of woodland at the site, and the deer are pseudoreplicates of the effect of woodland.

5.2.3 Relatedness

We all know that genetics mean that relatives tend to be more similar to each other than do unrelated members of a population. This similarity due to genetics means that relatives are not independent data points when looking at the effects of other treatments. So if we were looking at the effects of fertilizers on seed development, but all of the seeds in one treatment group came from one adult plant, while all the seeds in the other treatment came from another adult plant, these seeds would not be independent measures of the effect of the fertilizer treatment. Differences between the two groups of seeds might be due to the fertilizer treatment, but could equally be due to any genetic differences between the two groups. In the studies of chimps and deer discussed earlier, as well as the problems highlighted, there are also possible problems of relatedness between the individuals in each chimp enclosure or at each deer site.

5.2.4 A pseudoreplicated stimulus

Imagine that we were interested in whether female zebra finches prefer males with bright red beaks. We might carry out the following type of experiment. We take two stuffed male zebra finches and place them on opposite sides of a cage. One of the males has a bright beak, the other a dull beak. We then place 20 females in the cage one at a time, and see which model they spend most of the time next to. We find that on average, females spend more time with the bright male than the duller male. Can we now conclude that females prefer males with bright beaks? Again, the answer is 'no', and the reason is that our stimuli are pseudoreplicated. It is true that the stuffed finches differ in their beak colour, but they probably differ in any number of other ways as well, for example maybe one is slightly bigger than the other. The differences we observe in response to the two models might be due to their beak colours, but might be due to any of these other differences. Maybe your mount with the duller beak also smells unusual because of where it was kept prior to the experiment.

Q 5.1 Why in this mate-choice experiment do we test the females individually, rather than saving time by introducing them all into the choice chamber simultaneously?

5.2.5 Individuals are part of the environment too

A particular problem of behavioural studies is that individuals will affect each other. The chimps that we discussed previously are social animals, and the way in which one chimp behaves will have effects on the other chimps in the enclosure. So a single aggressive individual in a group might make all the others more aggressive. This can be thought of as an extension of the shared environment problem, but with the animals being the part of the environment that is shared. Thus, even if hypothetical identical cages or incubators really did exist, individuals within an enclosure will still share this part of the environment. The problem is not limited to behaviour either, but will arise whenever the state of one individual in a group can affect the state of others. With the deer and worms discussed earlier, it is quite possible that if one individual in a group has lots of worms it will infect all the other members of its group, and so the worm burdens of the members of a group are not independent.

5.2.6 Pseudoreplication of measurements through time

Imagine you are interested in the ranging behaviour of elk. Specifically, you want to know whether a particular elk individual prefers different types of habitat. To carry out your study you manage to get hold of some satellite tracking equipment that can tell you the position of the elk at specified time points. For each measurement that you make, you note down the type of vegetation that the elk is in. You find that of the 20 measurements that you made, 19 occurred in habitat that you classified as dense woodland, and only one occurred in open woodland. In your survey area as a whole there are equal amounts of these two habitat types, so does this evidence suggest that this elk prefers to spend time in dense woodland? The validity of this conclusion will depend on whether each of your measurements on an individual elk can be regarded as independent measures of the position of that elk. Whether or not this is true will depend (in part) on how much time has passed between measurements. Imagine first that we have taken the position of the elk once a day (at a random time of day) for 20 days, and that our study area is such that an elk could easily walk from any point in the site to any other point in the time between measurements. If that is the case then we have no reason to expect that the position of an elk at one measurement point will affect its position at any other measurement point. Contrast this with the same study, but with measurements taken every second for 20 consecutive seconds. In this (admittedly pretty extreme) case the position of an elk at one measurement point will obviously have an effect on where it is found at the next measurement (one second later). An elk cannot move far in a second, and so an elk that is found in dense woodland at one point is far more likely to also be in dense woodland a second later. If we had been silly enough to carry out the study in this way, we would not have independent data points, and our inference about habitat use would be flawed. Because of the non-independence of data points, even an elk with no habitat preference would still be highly likely to be found in the same habitat type every second for 20 seconds. Remember also that even if we collect good data of the ranging behaviour of this elk, it is only one individual, and it would be risky to assume that it is a representative individual of the wider population—we would want to replicate elk in order to draw conclusions about more than one individual.

Similar problems can occur in any experiment where we are taking multiple measurements through time, and whether or not measurements are independent will depend critically on the biology of the system. Even daily measurements might not be independent if we were studying a snail rather than an elk in the previous example. once again, you can see that deciding whether you have a pseudoreplication problem is a decision that involves understanding of the biology of the study system.

5.2.7 Species comparisons and pseudoreplication

It is often not possible to directly manipulate traits experimentally. However, as life scientists, we are fortunate to have a wide variety of species with different trait values that allow us to make comparisons. Imagine we hypothesize that in species where

Q 5.2 You are engaged in testing whether two factors vary between men and women: propensity to have red hair and willingness to engage in conversation with strangers. While waiting in the dentist's waiting room to be called through for a check-up you realize that the other occupants of the waiting room might provide an opportunity to add to your data set. How many independent data points can you collect for each of the two factors?

females mate with several males, sperm competition will lead to an increase in the relative testes size of the males. To test this, we take two species of damselfly, one where females mate once, and another where females mate multiply. We then compare the relative size of the males' testes, and find that those of the males in the promiscuous species are bigger for their body size than those of the monogamous species. Can we conclude that our hypothesis is correct? Well not really, because the measurements on the individuals of the two species are not really independent measurements of the effects of the mating systems. It is true that all the males of a species do experience the same mating system and this differs between the species, but they all experience a whole range of other factors that differ systematically between the species as well. The differences observed might be due to the factor that we are interested in, but might be due to any of the other factors that differ consistently between the two species. Maybe one species has a large body (and so relatively smaller testes) because it must carry large fat reserves to cope with an environment offering a more unpredictable food supply than that experienced by the other species. Thus multiple measurements of a species are not independent measures of any single difference between species.

So what about if we look at several species of damselfly and find that the pattern is consistent over several species? Can we now have confidence that the effect is due to mating system? Unfortunately the situation is not that simple. The dynamics of the evolutionary process mean that closely related species will tend to be similar to each other simply because of their shared evolutionary history. Thus, suppose we have ten species of promiscuous damselfly species, all with relatively large testes, but these have all evolved from a single ancestral species that was promiscuous and had large testes—we would have to be very cautious about drawing strong conclusions about our hypothesis based on our ten species. Furthermore, closely related species are likely to share many other aspects of their ecology too. There are statistical methods developed to deal with these problems of shared ancestry, and Harvey and Purvis (1991) in the Bibliography is a good place to start if this interests you.

It's easy to produce pseudoreplicates rather than independent replicates if you are unwary.

5.3 Dealing with non-independence

After Section 5.2, you must be thinking that non-independence of replicates is a minefield that will cause you no end of trouble. Not so! Forewarned is forearmed, and you will avoid lots of pitfalls just by bearing in mind the cautions of Section 5.2. However, this section will offer more practical advice, armed with which pseudoreplication should hold no fears.

5.3.1 Independence of replicates is a biological issue

The most difficult thing about the concept of independence of replicates is that there are no hard and fast rules. Instead, whether things are pseudoreplicates will depend on the biology of the species that you are studying and the questions that you are asking. *Thus, pseudoreplication is a problem that has to be addressed primarily by biologists and not by statisticians.* If you give a data set to a statistician, there is no way that they will be able to tell by looking at the data if your replicates are independent or not. A statistician might be able to guess that the behaviour of one chimp will affect the behaviour of others in the same social group, but can't be expected to know this; let alone whether the behaviours of beetles that are kept in separate pots inside an incubator are likely to affect each other. This is why biologists need to be involved in the design of their experiments. So use your biological knowledge to ask whether two measurements should be expected to be independent or not.

There are many solutions to pseudoreplication, and the course of action that you take to avoid it will depend on the particular situation that you are in. Sometimes the solution is straightforward: in the female-choice experiment with the pseudoreplicated stimulus, one solution might be to sample from a selection of stuffed males so that every female is tested with a different pair of stuffed males.

A more general solution involves condensing and replicating. In the example of the deer we decided that individual deer were not independent measures of woodland or moorland habitats. The first step is to condense all the measures of the deer from the woodland site to a single data point. This might be achieved by picking one at random, or by taking the mean for all the deer at the woodland site. We then do the same for the moorland site. However, this still leaves us with a problem, as we now have only one measurement for woodland and one for moorland. All the replication that we thought we had has vanished. The solution is then to go and find more woodlands and moorlands and do the same at each. This would give us several independent data points of deer from woodland sites and several of deer from moorland sites, and any differences we found could then be put down to more general effects of woodland or moorland. The same procedure could be used for the insects in the incubators (take a mean for the incubator, and replicate incubators), and the chimps in the enclosures (take a mean for the enclosures, and find more enclosures). Throwing away all those data by lumping them all into one mean value might seem like a terrible waste of effort. However, it would be far more of a waste of effort to pretend that the measurements were really independent and draw unreliable conclusions. It is worth asking at this point whether this implies that measuring several deer at each site was a complete waste of time. Would we have been just as well measuring only a single deer at each site, given that we ended up with a single measurement at each site? This is a topic that will be discussed in more detail in Section 11.2.1. Here we will simply say that because the mean of several deer at a site will be less affected by random variation than a single measurement, it will generally pay to measure several animals at a site, even if we ultimately condense these measurements of different animals into a single mean for the site. The extra work of collecting droppings from more deer at each site will give a more precise measure of that site and mean that fewer different sites will be required in your final analysis. As a life scientist, you'll probably enjoy looking for deer droppings more than driving between sites, so this is an attractive trade-off.

Sometimes it will not be practical to replicate in this way. Maybe the lab you are working in only has two incubators. Does this mean that your study of the effects of temperature on courtship song in crickets is doomed? There is another type of solution that is open to you in this situation. Imagine that we run the experiment with incubator A at 25 °C and incubator B at 30 °C. We can then repeat the experiment with incubator B at 25 °C and incubator A at 30 °C. We could do the same with the enclosures in the enrichment example, first enriching one enclosure, then the other. This means that we now have replicate measurements of crickets raised in incubator A at both temperatures and also in incubator B at both temperatures. By comparing the beetles at different temperatures in the same incubator, we have independent measures of temperature effects (although even this assumes that the incubators do not change between our runs of the experiment).

Lastly we should point out that there are also more complex statistical techniques that can handle non-independent replicates. In our deer example, we could enter every deer as a replicate in such a 'multi-level statistical model', but as well as recording the density of parasitic worms in the droppings for a particular deer we record the identity of the site that that particular deer comes from. Such statistical techniques can accommodate the assumption that deer that come from the same site (i.e. that share the same site identifier) may not be independent and control for this when testing the main hypothesis we are interested in (a difference between our woodland sites and our moorland sites). But these statistical techniques are more challenging to implement than simpler ones that we could use on independent data points (i.e. if we simply used the mean value for each site).

So your first approach should be to try and make sure all your replicates can be considered independent by even the Devil's Advocate (for example by using different model males for each female in our beak-colour example). If this is impossible, you should look to aggregate replicates into groupings that can be considered independent. In our example deer within a single site cannot really be considered independent, but different woodland sites miles from each other can be considered independent. Then we can include only one measure (usually the mean of a number of subsamples) for each independent replicate, and carry out a simple statistical test that assumes independence of replicates; or we can include all the subsamples (all the deer in our example) separately, but indicate which individuals are likely to be related in some way (sharing the same wood in our example) and carry out a more complex multi-level statistical test.

> You can almost always design your study to address the worst problems associated with non-independence. Any remaining non-independence can be handled by careful application of statistics that will avoid you pseudoreplicating.

5.4 Accepting that sometimes replication is impractical

Suppose that we really can only survey the deer on one woodland and one moorland, does this mean our study of the effect of woodland on parasitism is a waste of time? Not necessarily. The important thing is to ask yourself what the data tell you. Suppose we did find more worms in the droppings of woodland deer than in those of the

Q 5.3 You are carrying out a clinical trial of a drug that is reputed to affect red blood cell count. You have 60 volunteers that you have allocated to two groups of 30. One group is given the drug for a week, the other is not. You then take two blood samples from each person, and for each blood sample you examine three blood smears under a microscope and count the red blood cells. This gives you a total of 360 red blood cell counts with which to test your hypothesis. Would you be concerned about non-independence in this study? If so, how would you deal with this?

Q 5.4 Was measuring six smears from each individual in question 5.3 a waste of time?

moorland deer, what can we infer? Well we can say with some confidence that groups of deer differ in their worm burdens, but this is unlikely to be a very exciting result. If the deer at the two sites are similar (in species, density, age structure, etc.), then we can have some confidence that this is due to differences between the two sites rather than other differences between the two groups of deer. However, we can't say that this is due to the fact that one site is woodland and one site moorland, because of all the other differences between the two sites. So what began as a study of an interesting question 'do moorland deer have fewer worms than woodland deer?' has been limited to answering the question, 'do all groups of deer have the same worm burdens?' This is typical of what happens if we don't think carefully about our replicates—we think we are answering an interesting question, but in fact we can only answer a more mundane question to which we probably already know the answer.

Replicates allow us to answer general questions and the generality of the answer that we get will depend critically on how we sample and replicate. If we are interested in the effect of sex on the heights of humans, multiple measurements of Pierre and Marie Curie are obviously of vanishingly little use. They tell us about differences between Pierre and Marie, and if we can only measure their heights with a great deal of error we might need multiple measurements even to answer such a simple question. However, these multiple measurements will tell us very little about sex differences in humans more generally. If we measure the heights of several males and females then, as long as we have avoided traps of relatedness, shared environments (did all the males come from one country and all the females from another?), and so on, we can begin to say something more general about the effect of sex on height. However, even if we have sampled carefully, if all the individuals come from Britain they are independent measures of the heights of British males and females, but not of males and females at the global level. So if we were interested in knowing the answer to this question at a global level we would need a sample of independent replicates drawn from the global population.

➡ The key if you can't replicate fully is to be aware of the limitations of what you can conclude from your data. Pseudoreplicated enclosures mean we can talk about whether there are differences between the enclosures, but we can't be sure what causes any differences, and pseudoreplicated stimuli mean we can talk about the effects of the stuffed models on female choice but can't be sure these effects are due to the beak colour that we were originally interested in.

5.5 Pseudoreplication, third variables, and confounding variables

In Chapter 3 we introduced the problems of third variables in relation to correlational studies and in this chapter and the last we have talked about replication, pseudoreplication, and confounding variables. Here we will discuss how these ideas are linked.

As we have already said, if we want to know the effect of some experimental treatment, say the effect of an antibiotic on bacterial growth rates, we need independent replicate measures of the effect of that treatment. That is, we need several measures of the growth rate of bacteria with and without the antibiotic, and the only systematic difference between the bacteria in the treatment and control groups must be the presence of the antibiotic. If the bacteria in the two groups differ systematically in other ways as well, then our replicates are actually pseudoreplicates. For example, imagine we took two agar plates containing bacterial growth media, and added antibiotic to one of the plates. We then spread a dilute solution of bacteria on each plate, and then measured the diameter of 100 bacterial colonies on each plate after 24 hours. We appear to have 100 measures of growth rate with and without antibiotic, but of course these are not independent measures, because all the bacteria in a group are grown on the same plate. The plates differ with respect to antibiotic application, but they will also differ in other ways as well; the differences due to antibiotic will be confounded with these other differences. Or to put this in the language of Chapter 3, we cannot be sure that the difference we see is due only to the factor that we are interested in (wholly or in part) or to some unmeasured third variables. So confounding factors and third variables are the same thing, and if they are not dealt with appropriately they will turn our replicates into pseudoreplicates. One solution to the problem in the bacterial study might be to have several plates with and without antibiotic, and then measure the growth rate of a single colony on each plate (or of several colonies and then take a mean for each plate).

➡️ The problems of third variables in correlational studies can equally be thought of as a problem of pseudoreplication. Because we have not manipulated experimentally, we cannot be sure that the individuals that we are comparing really are replicate measures of the effects of the factor that we care about, rather than being pseudoreplicates that are confounded by other factors.

5.6 Cohort effects, confounding variables, and cross-sectional studies

Another area where non-independence and confounding effects can trip up the unwary is when we are interested in changes that occur through time. So, imagine we find that a sample of 40-year-old men had better hand–eye coordination in a test situation than a sample of 20-year-old men? Can we safely assume that men's coordination improves as they get older?

In a **cross-sectional study** we take a snapshot of a population at a single moment in time and compare between individuals of different ages or developmental stages.

Our study is **cross-sectional**, where we take a snapshot in time and compare subjects of different ages. If we find a difference between different age classes then this could indicate that individuals do change over time, but there is an alternative explanation. It could be that individuals do not change over time but that there is a difference between the individuals in the different age classes due to some other factor. For example, the types of jobs that men typically do have changed dramatically over the last 20 years. Perhaps a greater fraction of the 40-year-olds have jobs that involve manual labour or craftsmanship, whereas more of the 20-year-olds have office jobs. If the type of job that a person does affects their coordination, then this difference in employment may explain the difference between the groups. You will recognize this immediately as another example of a confounding factor; although in studies comparing different age-classes it generally goes by the name of a *cohort effect*. An alternative way to probe the question of whether an individual's coordination changes with age is to follow individuals over time and compare performance of the same individuals at different ages: this would be called a **longitudinal study**. But, hang on, given what we have said about multiple measurements on the same individual not being independent, alarm bells should now be ringing. Surely, by measuring the same individuals at different ages we are doing just that? For the time being you will just have to take our word for it that in situations like this where we are specifically interested in how individuals change with time or experience, then (as long as the data are handled with care) this is not a problem. The details of why it is not a problem can be found when we discuss within-subject designs in Chapter 10. Thus, a longitudinal study avoids the potential for being misled by cohort effects, but obviously takes longer to provide us with an answer than a *cross-sectional study* (where we take a snap-shot and compare subjects of different age). Hence a reasonable compromise is often to take a combination of the two. As another example, imagine that we wanted to explore fecundity in female lions to see if it declines after a certain age. If a cross-sectional study showed that three-year-old lions had unusually low fecundity, this might not be anything to do with being aged three, it could be that food was particularly scarce three years before when these lions were cubs, resulting in that age cohort being generally of poor quality (this is a cohort effect). We can test for this cohort effect: if we repeated the study next year, we would expect to find that now four-year-olds would have low fecundity if we have a cohort effect. Alternatively, if the second study shows the same age-effect as the first, then this suggests that the low fecundity of three-year-olds is a more general developmental effect.

> In a **longitudinal study** we follow individuals over time and compare the same individual at different ages or developmental stages.

 If you are looking to detect developmental changes, then be on your guard for cohort effects.

If you want to test your understanding of the concepts introduced in this chapter a little more, then there are some additional self-test questions in the Supplementary Information. Otherwise, we can move on to the next chapter where we focus on how large you might want to make your study (i.e. how many independent subjects you might want to collect data on).

■ Summary

- In the life sciences, we are almost always working with a sample of experimental subjects, drawn from a larger population. To be independent, a measurement made on one subject should not provide any useful information about the measurement of that factor on another subject in the sample.

- If your sample subjects are not independent, then your statistical treatment of your data could be misleading if you treat your subjects as if they were independent.
- Strive in your experimental design to make sure you can argue that subjects (or groups of subjects) are independent in a way that will satisfy the Devil's Advocate.
- Non-independence is an issue about biology, it requires biological insight; you cannot turn to a statistician for salvation, but must face it yourself.
- The key if you can't replicate fully is to be aware of the limitations of what you can conclude from your data.
- The problems of third variables in correlation studies can equally be thought of as a problem of pseudoreplication and confounding factors.
- If you are looking to detect developmental changes, then be on your guard for cohort effects.

Sample size, power, and efficient design

6

Having decided on the specific hypothesis you want to test, and the broad outline of how you plan to design a study to collect data to test that hypothesis, a key question is how much data to collect. The more data you have, the more definitively you should be able to test your hypothesis, but there will be diminishing returns on ever greater effort in data collection. This chapter should help you decide on a sensible sample size; but we also emphasize that there is a lot more you can do to design efficient experiments.

- We begin by introducing the concept of statistical power as a tool to help you design efficient experiments (Section 6.1).

- We then discuss the factors that affect the statistical power of a study (Section 6.2), and how to estimate the power of a planned experiment (Section 6.3).

- We finish by exploring how the power of a study can be optimized (Section 6.4) and how we can compare different designs in terms of their power (Section 6.5).

6.1 Selecting the appropriate number of replicates

As we discussed in the last two chapters, replication is the basis of all experimental design, and a natural question that arises when planning any study is how many replicates do we need? As we saw earlier, the more replicates we have, the more confident we can be that differences between groups are real and not simply due to chance effects. So all things being equal, we want as many replicates as possible. However, as with all things in life, all things are not equal. Increasing replication incurs costs. These costs might be financial—if an experiment involves expensive chemical reagents then doubling the number of replicates will result in a large increase in cost. More likely (and more importantly for many people), experiments will involve time costs. Probably most important of all, if experiments involve using humans or animals (or materials obtained from humans or animals) then there may well be welfare or conservation costs of increasing sample sizes.

These costs mean that the general answer to the question of how big should my experiment be is that it should be big enough to give you confidence that you will be able to detect any biologically meaningful effects that exist, but not so big that some sampling was unnecessary. How can we decide how big this actually is? There are two general approaches that can be used to answer this question; we can either make educated guesses based on similar studies or carry out formal power analysis.

 'The more samples, the better' is an over-simplification.

6.1.1 Educated guesswork

The simplest way to get an idea of how big your experiment should be is to look at similar experiments that have been done by other people. This may seem obvious, but it is an extremely useful method for getting a ballpark figure for how big an experiment might need to be. It is also an extremely useful step in obtaining some of the biological data that you will need if you plan to go on and do a more formal power analysis. Never convince yourself that no similar experiments have ever been done before; there are *always* experiments already published similar enough to your own to be useful. This approach is simple, but very effective. We have only devoted one paragraph to it because it's a straightforward technique with no hidden pitfalls, not because we don't think it's important. We do think it is important: *go to the library*!

 Learn all you can from previous studies.

6.1.2 Formal power analysis

Statistical power is an extremely simple and useful concept that can be used to guide us when we are trying to decide how big an experiment needs to be. So what is it? Statistical power is the probability that a particular experiment will detect an effect (e.g. a difference between groups or a relationship between variables) assuming that there really is an effect to be detected. If this definition is a little terse for you then try the example in Box 6.1; otherwise, let's move on.

6.2 Factors affecting the power of an experiment

The power of an experiment of specific design will be affected by three main things: the *effect size*, the amount of *random variation*, and the number of replicates.

In the example in Box 6.1, the effect size is the real difference between the mean thickness of the two populations of egg shells, those from hens fed GM feed and those fed standard feed. In other cases the effect size might be the strength of the relationship between two variables. In fact, the effect size can be anything that we are trying to detect. The important thing about the effect size is that, all other things being equal, the bigger the effect size, the easier it will be to detect, and so the more powerful

BOX 6.1 An example of statistical power calculation: returning to hens' eggs

In Chapter 4 we used the example of the effects of GM feed on egg shell thickness to think about variation and replication (see Box 4.1). Let's return to this example to now think about statistical power. For the sake of this example we are going to assume that the feed containing the GM grain really does have an effect on the thickness of egg shells, causing an average reduction of 0.4 mm. Or to put it in terms that a statistician might use, the population mean of GM eggs is different from the population mean of standard eggs. Of course, our investigator does not know this, and is engaged in their usual quest to determine whether GM feed affects shell thickness. To do this they have measured five GM eggs and five standard eggs. Unsurprisingly, the average thicknesses of the two groups are not identical (go back to Box 4.1 if this does surprise you) and the investigator carries out an appropriate statistical test to examine whether the difference between the two samples that they have collected is big enough to suggest that GM feed really does have an adverse effect. In statistical jargon, the investigator wants to know whether these groups are significantly different from one another, suggesting that the populations that they represent have different means. If the investigator finds a significant difference, then they conclude that GM feed affects shell thickness; if they find no significant difference, they conclude that there is no evidence that GM feed affects shell thickness.

Let us imagine that the investigator gets a statistically significant result, and comes to the conclusion that GM feed has an effect on shell thickness. However, the researcher does not stop there. Instead, being very thorough, they come back the next day and repeat the experiment in exactly the same way as they did the day before. The shell measurements that they collect in this second experiment will differ from those of the day before, because of the effects of random variation (there are likely to be many factors that affect the thickness of an egg shell). This in turn will affect the mean size of the two groups. Again, the investigator does their statistics and finds a significant difference between the groups. However, even this is not enough for our intrepid researcher, and they are back the following day doing exactly the same experiment again. Again, the exact measures differ, and this affects the means. However, today (by chance) the five GM shells all happen to be slightly thicker than the average of their population due to random variation, while the standard shells are all thinner than would be expected, having negative deviations. This means that there is very little difference between the means of the two sample groups, and when the investigator does the statistics he finds no significant difference between the groups, so today he finds no evidence that GM feed has any effect.

This is a strange set of circumstances, so let's get this clear. The investigator has done nothing wrong in this third experiment, the hens were all randomly chosen, and the experiment was carried out correctly. The fact that the GM shells are all slightly thicker than average is just down to chance, like tossing a coin five times and getting five heads. It's unlikely to happen, but is always a possibility and that means it will happen sometimes. When it does, it will lead our researcher to draw the wrong conclusions and say that feed has no effect, when in fact it does.

However, at this point our investigator becomes very worried, and comes back every day for 100 days and repeats the experiment (we don't recommend you try this yourself). On 72 days he finds a statistically significant difference between samples, and on the other 28 days no evidence of a difference. Now remember, we know the correct answer: there really is a difference between the two populations from which all the samples were drawn. So, with this particular experimental design, the experimenter has got the correct answer on 72 out of 100 tries. Thus, the power of the experiment (the probability that it detects the difference between the populations when there really is one to detect) is about 72%.

the experiment will be. The reason for this is obvious—if a change in feed affects shell thickness by 20%, then this average difference between groups is less likely to be masked by noise than a 2% difference.

The **effect size** is the magnitude of the effect we are measuring. Effect sizes can include the difference between the means of two groups, or the slope of the relationship between two variables.

The **effect size** is the magnitude of the effect of one factor on the variable that we are measuring. If we are looking at the factors that influence a dog's weight then both breed and the amount of daily exercise might be expected to influence weight, but we would expect breed to have a stronger effect. If this is true, then the effect size associated with breed is higher than that associated with exercise.

An increase in the amount of **random variation** between individuals has the opposite effect to an increase in effect size. As the amount of random variation increases, it becomes harder to detect an effect of given size. In the hen example, we can think of the thickness of an egg as being partly due to the treatment group that the hen is in, and partly due to other factors. As the importance of the other factors increases, the effect of the treatment becomes harder to detect.

Random variation is the variation between sample units that cannot be accounted for by the factors considered and so is due to other (random) factors.

Finally, the size of the samples will affect the power, with more replicates increasing the power of an experiment. Again, the reasons for this will be obvious after our discussion earlier about the effect of replication. Replication acts to cancel out some of the effects of random variation, because with several measurements the chance of them all suffering similarly from random variation is very small. The more replicates we use, the more effective this cancellation will be. This reduction in the effect of within-sample variation will make differences between samples easier to detect.

6.3 Working out the power of a planned study

Statistical power depends in part on what the world is really like (i.e. how big the biological effects are and how variable the world is), and in part on how we plan do the experiment (i.e. how big our experiment is, and the particular design that we use). If we can make an educated guess as to what the world is like, then we can determine the power of different types of experiments. Another thought experiment should make this clearer.

Back at the animal feed company, our researcher is planning their next experiment to investigate whether adding a new food supplement, Factor Q, to their chicken feed will increase the mass of chickens. To ensure that they design their experiment sensibly, they decide to use power analysis to help them decide on the appropriate number of chickens to use in their study. There are four key factors in any experiment that will have an effect on its power:

1. The effect size (e.g. how big the real difference is between the groups, or how steep the relationship is between two variables).
2. The amount of random variation in the thing we are measuring.
3. The experimental design (and statistical test).
4. The number of independent replicates.

The first two factors relate to the biology of the system, or what the world is like. The effect size in this case is the effect that the food supplement has on growth, or in the words of a statistician, the difference between the means of the population of chickens fed the new supplement and the population of chickens fed the standard feed. The random variation will be due to all of the other things that affect a chicken's growth rate. The researcher will have limited control over these factors, although they can minimize the amount of random variation through careful experimental procedure. More importantly, the researcher will not know exactly what the values for these factors are (that's why they are carrying out the study), and so will have to make some educated guesses. Factors 3 and 4 are in the control of the researcher.

6.3.1 Determining the likely effect size

All other things being equal, the bigger the effect of our treatment, the easier it will be to detect; and so power generally increases with effect size. Thus, to determine the power of a study the researcher needs to decide what effect size they are trying to detect. Of course, the researcher will not know in advance how much of an effect the treatment will have, otherwise they would not need to do the experiment at all, and so instead they have to make an educated guess. Now this guess might be based on results for other similar supplements: maybe other supplements tend to increase weight by around 500 g, so the investigator expects a similar effect from their supplement. The experimenter might have carried out a small pilot study to get some idea of sizes of effects before embarking on the full experiment. Alternatively the guess might be based on theoretical predictions. Maybe there are physiological models that suggest that this type of supplement would be expected to increase weight by 500 g. Finally, maybe there are reasons that the investigator would not be interested in an effect of below a certain size. For example, maybe a supplement that increased weight by less than 500 g would have no use in the poultry industry, so the investigator is only worried about being able to detect an effect of at least 500 g. Whatever the reason, our researcher

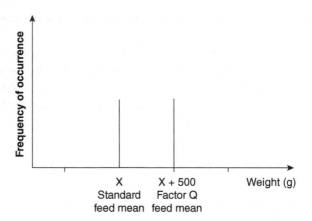

Figure 6.1 Building a picture of what the world might be like: step 1. The two populations that our experiment will sample from have different means since the null hypothesis is false.

decides on 500 g as the effect size that they want to be able to detect in their experiment. In Box 4.1, we illustrated how experiments can be thought of as drawing random samples from populations. Now that our researcher has decided on an effect size they can begin to draw a picture of what they think these populations look like. Figure 6.1 shows what this picture looks like, with two populations, one for chickens fed standard feed which has a mean of some value X, and one for chickens fed factor Q which has a mean of X + 500.

6.3.2 How much variation?

Of course, even chickens kept in identical conditions and fed the same food will vary in weight due to all the other factors that influence chicken weight, and our researcher's next task is to think about what this variation will look like and how much there is likely to be. Again the researcher will not know exactly how much the sizes of chickens in their two groups will vary due to random factors, and so they will need to make a second educated guess. As with the effect size, there are numerous ways of doing this, but the easiest is to use biological knowledge of the system (or other similar systems). In a study based on commercial animal production there is likely to be a great deal of information already available. The researcher might look at other studies of chicken weight in similar experimental settings and base their estimate on these. If our researcher had been working on a much less intensively studied system they might decide to carry out a small pilot study to get an estimate of variation in chicken weight before embarking on their main study. In this case our researcher knows that the body weight of chickens in his facility tends to be normally distributed with a SD of some value Y g, and decides to use this as their estimate of random variation. The researcher can then add this random variation to their picture of the world and this is shown in Figure 6.2. Now this picture shows not only that our two populations have different means, but also the dis-

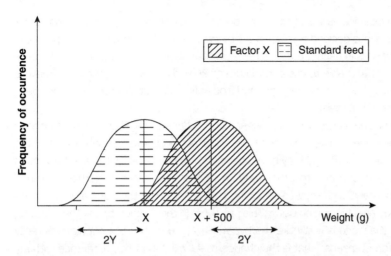

Figure 6.2 Building a picture of what the world might be like: step 2. We can now add variation to our picture of the world. We assume that the variation follows a Gaussian distribution, and that the SD of each distribution is Y.

tribution of actual sizes that we expect to see due to random variation. The researcher can then begin to ask, if the world is really like this picture, whether their experiment is likely to detect the effect of the treatment.

6.3.3 Experimental design

In some situations there may be different ways in which a particular study could be carried out, and the particular design used (and the statistical analysis that this dictates) may have an effect on the power of the study. Questions such as how many treatment groups to have, what levels of treatment to apply, and whether to have equal numbers of replicates in each treatment group all need to be considered. These issues will be discussed in more detail in Chapter 7, and we will not dwell on them here. In this case, our researcher has decided on a very simple design: two equally sized groups of chickens, one group fed the normal diet and one fed the normal diet plus the supplement.

6.3.4 How many replicates?

The final factor that will affect power is the number of replicates that the investigator is planning to use. Indeed the most common use of power analysis is to allow researchers to compare the power of experiments with different numbers of replicates and so decide on an appropriate level of replication for their study. As we have discussed before, the ideal number of replicates will be big enough to detect the effect that we are looking for, but no bigger. Of course, the actual number of replicates will also be affected by practical considerations as well: if we can only fit 20 chickens into our research set-up then this sets the maximum size of our experiment, but power analysis allows us to examine how best to use resources within these constraints. In this case, our researcher has decided they will start by considering the power of an experiment with five chickens in each group. Further, power analysis could indicate that the constraint of only being able to use a maximum of 20 birds means that the experiment is not worth doing because it is very unlikely to be able to detect differences of the specified size, even if they exist.

OK, let's just summarize what our researcher has done so far. First they have defined a world where the supplement does have an effect on chicken growth rate (i.e. the null hypothesis is false), and they have specified how big this effect is, and how much chickens vary in size due to other random factors. The researcher then specified the experimental design they are planning to use, including the number of replicate chickens they are planning on using. This put them in the position to ask the following question. If the world is really like the hypothetical world, and the populations really do have different means, what is the chance that I will detect this difference with my experimental design?

6.3.5 Imaginary experiments

In Box 6.1 our researcher estimated the power of their particular experimental design directly, by actually carrying out the experiment many times, and seeing how many of those experiments detected a difference. For obvious reasons, this would be an impractical way to estimate the power of a proposed study in practice. However, our researcher can use a computer with a random number generator to carry out imaginary experiments in a similar way to that seen in Box 6.1. The process is outlined in Figure 6.3.

First they generate five imaginary chickens fed on the standard diet. Remember, you can think of experimental subjects as being a random sample drawn from a large population. In their hypothetical world, the population of chickens fed standard feed has a mean of 1500 g and a SD of 300 g, and so the researcher gets the computer to generate five random numbers drawn from a normal distribution with a mean of 1500 and a SD of 300. They then generate five imaginary chickens fed on supplemented feed in the same way, except that these chickens are drawn from a distribution with a mean of 2000 g and a SD of 300 g. The researcher has now collected imaginary data of exactly the sort they are planning to collect in their real experiment, and so they can carry out the same test they are planning to use to see whether they find a significant difference between the groups in their imaginary experiment. Of course, the researcher knows that for this imaginary experiment the null hypothesis is false, because they have set up the computer-model world that way, so if they obtain a significant difference, their imaginary experiment has drawn the correct conclusion, whilst if the difference between the groups is not significant they would draw the wrong conclusions from their experiment.

You should be familiar enough with variation by now to see that simply doing this once will not reveal what the researcher wants to know. The conclusions of their imaginary experiment will depend on the particular random sample of chickens that the computer generated for them. If they repeated the experiment, they would almost certainly get

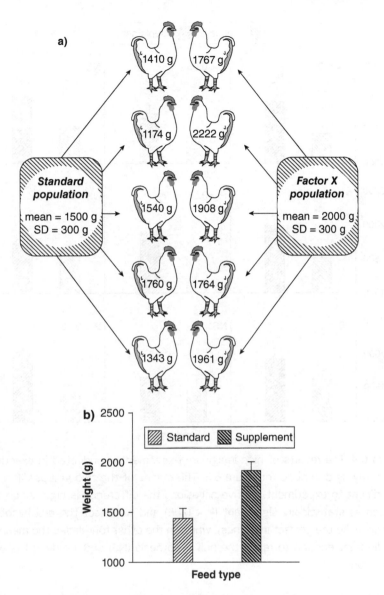

Figure 6.3 Estimating the power of an experiment by simulations. We define the characteristics of the populations we are sampling from, and simulate the experiment by drawing the required number of weight measures from our populations. We then use statistical software to examine whether the means of our populations are significantly different. By repeating this process several thousand times we can examine how often our design detects the difference between the populations that we have defined.

different weights for each chicken, and might draw different conclusions just as we saw in the real experiments in Box 4.1. Indeed Figure 6.4 shows just that: when we repeat the experiment nine times, the actual means for the groups vary from trial to trial, and sometimes they are statistically significant, sometimes not.

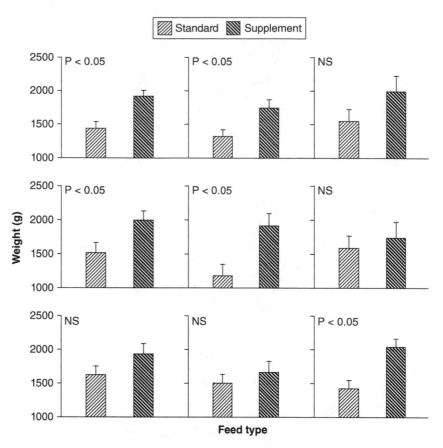

Figure 6.4 The results of nine imaginary experiments conducted in exactly the same way as described for Figure 6.3. The means of the two groups differ from experiment to experiment. On five occasions the difference is big enough to be classed as statistically significant ($P < 0.05$) and we reject the null hypothesis (and so make the correct inference), whilst in the other four cases the means are not different enough to reject the null hypothesis (NS) and we draw the wrong conclusions.

To say anything general about the power of the design, the researcher needs to repeat this process many, many times. However, the great thing about doing this imaginary experiment on a computer, rather than with real chickens, is that computers are very good at doing the same thing over and over again. In this case, the researcher repeats the procedure 1000 times; in some of the imaginary experiments, the difference between the groups is detected as significant and the experiment has generated the correct answer, and in others no significant difference is found. In total, the researcher finds a significant difference in 503 of the 1000 simulations, giving their experiment a power of about 50%. In practical terms, this means that if their view of the world is correct, and they carried out the experiment with five chickens in each group, they would only have a 50:50 chance of detecting any difference caused by the supplement.

6.4 Improving the power of a study

Inevitably there will be times when your power analysis indicates that your planned study will have low power. Thus, suppose that you are interested in whether a proposed new food additive causes anaemia, and are testing this using laboratory mice. Your initial plan is to feed 10 mice with standard food, and feed another 10 mice food containing the additive. You will then estimate their level of anaemia by measuring their red blood cell count after one week on the diets. However, when you carry out your power analysis you are disappointed to see that the predicted power is only 23%. To proceed with the experiment, when you have less than a one in four chance of detecting the kinds of effects that you are interested in even if they exist, would clearly be a bad idea. It would be a waste of time and resources, and maybe an irresponsible use of animals.

So what can you do? Based on the discussions in this chapter so far, the first solution that will probably spring to mind is to increase the size of the study. Perhaps if you used 20 or 30 mice in each group you could achieve a reasonable power. However, as we have already emphasized, increasing sample size without limit is generally neither possible nor desirable. Mice are expensive to buy, house, and maintain, and you are unlikely to have an unlimited budget. Similarly, since the ever extending mouse house has not yet been built, there will be practical constraints on how many mice you could work with at one time. And of course, mice are sentient creatures and so simply increasing the numbers we use in an experiment, without considering alternative strategies, would be ethically dubious. Fortunately, armed with the knowledge of how the different aspects of a study affect its power we can consider alternative options.

Q 6.1 You are interested in whether fever has a protective effect on mice infected with malaria and are planning a study in which mice that have been infected with malaria are either given an anti-inflammatory drug that reduces their fever or a placebo. Figure 6.5 shows power curves for two different sizes of experiment. Which should you use?

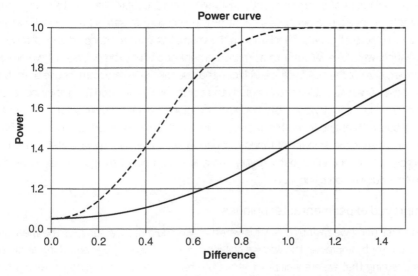

Figure 6.5 Power curves for experiments with 5 mice per group (solid line) and 25 mice per group (dashed line). The x-axis shows the expected effect size (the difference in average temperature between control and treated mice), whilst the y-axis shows the predicted power.

Before we consider these options, you are probably wondering what constitutes an acceptable level of power in your experiment. As a starting point we think 0.8 is a good number to have in your head. That means if we do the experiment, and effects of the size we are interested in exist, then we have an 80% chance of picking them up. These are pretty good odds, but you might wonder why we don't strive for a higher value, like 95%? There is a law of diminishing returns with statistical power: the greater your power already is, the more you have to improve your experiment to get the same incremental improvement in power. So it is likely that an experiment with 95% power would have to be much, much bigger than one with 80% power. Bigger means more expensive, more time consuming, and perhaps more demanding of animals. Hence, there are practical and often ethical reasons why it does not make sense to strive for ever more power. But there is nothing sacred about our 80% guideline—there might be times when a little higher or lower power makes sense (but be prepared to have a justification ready for the Devil's Advocate).

Q 6.2 A colleague appears and excitedly tells you that they have completed their power analysis, and their predicted power is 96%. What advice might you give them if their project involves measuring antigen levels in captive primates, or measuring leaf damage on individual oak leaves on local trees?

6.4.1 Reducing random variation

Random variation between samples gets in the way of our ability to see effects in our studies, and the more random variation there is, the lower the power of our study. In the case of our mouse anaemia study, the random variation is the variation in the red blood cell counts we measure that is due to anything other than our experimental treatment. It will include things that affect the average red blood cell count of each mouse: for example, age, general health, gender, or genotype may all affect red blood cell count and may vary amongst the individuals in our study. Similarly, we probably don't know the red blood cell count of our mice exactly. Instead we are probably planning to take a small sample of their blood, and use a count of the cells in that sample as our estimate of the red blood cell count of each mouse. This sampling can introduce additional random variation, which we will refer to as *measurement error*. For example, even if we are extremely careful the exact volume of blood that we take may vary from mouse to mouse, and even if we were able to take exactly the same volume of blood from our mouse, the red blood cell count would vary because red blood cells are not absolutely uniformly distributed throughout the blood plasma. In our discussion of power so far, we have assumed that the amount of random variation amongst our experimental subjects is largely out of our control, but in reality, there are a number of things we may be able to do to reduce random variation.

Improved experimental techniques

The first thing we might consider is whether we could reduce some of the measurement error by improving the way we collect our data. This may be something as simple as ensuring that we are well practised in the techniques we are going to use, so that we are not introducing unnecessary noise. Or perhaps we should consider whether the method we are using to estimate red blood cell count is as precise as it could be—perhaps an automated cell counter would allow us to examine larger samples of blood and so get more precise estimates than we can get by counting cells down a microscope?

We should always keep in mind that there may be other measures that we could make that will still answer our question, but are less variable. All of these are questions you should be asking as part of the pilot study that we have encouraged you to carry out throughout the book. The bottom line is that we should be doing all that we can to reduce measurement error, as doing so will reap benefits in terms of increased power. In Chapter 11 we will discuss in greater depth ways to ensure that this is the case.

More uniform experimental conditions

Once we are happy we have done all that we can to reduce measurement error, we can start to think about how we might reduce other sorts of random variation amongst our samples. Obviously, anything that we can do to standardize conditions across our study will help to reduce variation. This is exactly the reason that many studies are best carried out under carefully controlled laboratory conditions. In our case we would want to ensure that our mouse cages are as similar as possible with, for example, standardized food regimes. We would also want the conditions in the animal house where we are conducting our study to be as uniform as possible. If the facility has several rooms, then we would aim to have our study carried out in the one room, and if each room has several racks for cages, we would ask to have our mice housed in the same rack (or, if several racks are needed, in racks close together).

More uniform experimental material

Anything that we can do to improve the standardization of our experimental conditions should help to reduce random variation. However, we may also know from previous studies about specific factors that affect the variation amongst our mice. In that case another strategy that we might consider is to limit our study to a subset of individuals with a particular value for these factors. For example, if we know that age affects red blood cell count in mice we might choose to only use mice of a particular age, reducing or removing this particular source of variation.

For many characters in many species an incredibly important source of variation between individuals will be due to the genes that they carry. If we are lucky enough to be working with a well-established experimental model organism like a mouse, we may even be able to do something about that source of variation by making use of inbred lines. These are lines of genetically uniform individuals, with names like BALB/c and C57BL/HeN, that have been created thorough generations of selective breeding. As you would expect, these genetically uniform lines often show considerably less variation between individuals than you would see if you took individuals from a more genetically diverse population. By working with a single inbred line we are effectively choosing to work on a single genotype in the same way as we might choose to work on individuals of a single age if we expect age to affect the thing we are measuring.

If we do decide to limit our study to a more uniform subset of the population that we are interested in, there is something important to keep in mind. In Chapter 4 we emphasized that when we select individuals or samples for our studies, we want these to be representative of the population that we are interested in. If they are not, then the Devil's Advocate may question the external validity of our results. If we choose to

Q 6.3 From the perspective of statistical power, what are the advantages of carrying out experiments on organisms such as *Daphnia* (water fleas) or *Lemna* (duckweed) that can reproduce clonally?

work with a more limited subset of that population, say three-months-old male BALB/c mice, then technically any results of our study should only be applied to the population that they represent (other three-months-old BALB/c mice). Of course, how far any results can be generalized outside of the specific population studied depends on the biology of the system. If we find that the food additive is harmful to three-months-old BALB/c males, then even the most cynical critic would probably be happy to accept that it is very likely to be harmful to three-and-a-half-months-old BALB/c males, and probably to all adult males of that genotype. Whether it is also likely to apply to females or other genotypes is a matter of biological judgement and interpretation. At first sight, you might find such advice worrying. What use is a study that only applies to a subset of the population? In fact, biologists make these kinds of generalization beyond the specifics of their study all of the time. We are usually happy to assume that a study done on a random sample of British males may often tell us something useful about male humans in general, or that a substance that induces tumours in rabbits may well indicate that it would do so in humans. The key is to remember that such extrapolation is an assumption of yours. You should be ready to persuade the Devil's Advocate (based on general biological knowledge) that the individuals that you have chosen to study can reasonably be expected to tell you something about the population that you really care about. But in general be aware that there might be a trade-off between narrowing your pool of experimental subjects to increase standardization and improve power, and your ability to generalize from that study.

 Q 6.4 You are planning a study to measure growth rates in bacterial cultures experiencing different environmental stresses. You are considering increasing the number of cultures that you measure, but this would mean your experiment would need to spread into a second incubator. What effect would you expect these changes to have on the power of your study?

There may also sometimes be a trade-off between increasing sample size in order to improve power and the improvement to power through using more uniform experimental material. That is, sometimes increasing your sample size can only be achieved by increasing between-subject variability. This might occur because increasing your sample size requires you to use domestic chicks from two different suppliers rather than one; or two different incubators for your petri dishes instead of one; or measure your plants in three different batches rather than all in the same session.

Anything we can do to reduce variation amongst our samples will increase the power of our study.

6.4.2 Dealing with variation through design

An alternative approach to dealing with defined sources of variation is to measure them and explicitly include them in our study. This might involve something as simple as measuring the additional factors for the individuals used in our study so that they can be included in our analysis. In the case of our food supplement study, we might decide that rather than only carrying out the study on three-months-old mice we will use mice that differ in age, but we will keep a note of the age. A more complex procedure might involve classifying all individuals into a number of age categories, such as young, middle-, and old-aged, prior to allocating them to treatment groups, and then ensuring that half of the individuals in each category end up in the two food treatments. These (and several other) designs will be discussed in much more detail in Chapter 9, so we will not go into further details here. How do these approaches improve the power of the proposed study? The short answer is that they allow us to use statistics to remove the variation due to these other factors, and less variation means more power. For a longer answer, see Chapter 9.

 Some sources of variation can be included in the design of the experiment and then removed with statistical techniques.

6.4.3 Increasing the effect size

Usually the effect size of a study is the aspect that you have least control over. After all, the effect size is determined by the biology of the system, and isn't that what we are trying to learn about? However, sometimes there may be ways to influence effect sizes through the levels of a treatment that we choose for our study. Take for example our study of additives and anaemia. If the additive does cause anaemia, it is likely that the degree of anaemia will depend on the concentration of the additive. Thus, if we use high concentrations of additive in our experimental treatment we are likely to generate a large effect (if the additive has an effect), and so our experiment will have higher power than if we used lower doses (see Box 6.2 for more consideration of this). Obviously, if we use a dose that is higher than the level than the additive would be normally used then a critic may question whether any harm that we detect is biologically relevant. As is so often the case, the answer to this question is biological rather than statistical.

Q 6.5 You are planning to examine how the tail length of male widowbirds affects their attractiveness to females by experimentally increasing and decreasing the lengths of their tails. What are the attractions and drawbacks of manipulating such that your long tail males have tails that are 5 cm longer than the longest tail that you have observed in the population?

 Increasing effect size increases statistical power.

BOX 6.2 How to select the levels for a treatment

Suppose that you want to know what effect a fertilizer has on the growth rate of tomato plants, and you have 30 plants at your disposal. How should you spread these plants across fertilizer treatments? One extreme would be to choose the highest level of fertilizer you might reasonably add (say 30 g) and the lowest (maybe 0 g), and run an experiment with just these two treatments, with 15 plants in each treatment. At the other extreme, you might decide to keep the same limits but have many more treatment groups spread between these values. You might choose to give one plant 0 g, one plant 1 g, one plant 2 g, and so on, increasing the dose in 1-g intervals until you get to 29 g. Or you could choose to have two plants at each concentration, and have treatments spaced in 2-g units. In fact, there are many ways to carry out this experiment—how do you choose which is best?

The answer to this question, as with so many questions of design, will depend on your knowledge of the biology underlying the system. Suppose that you know, or at least are very confident, that the effect of fertilizer on growth rate increases linearly with concentration (as in Figure 6.6a), and you simply want to know how

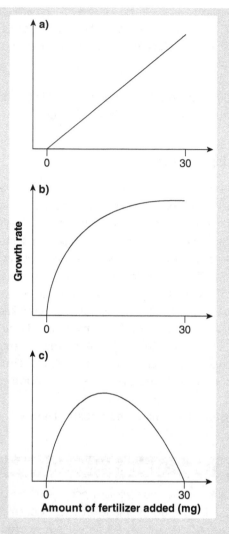

Figure 6.6 Three possible shapes for the relationship between amount of fertilizer applied and plant growth rate. In (a) there is a linear increase in growth rate with increasingly higher amounts of added fertilizer; in (b) there is also an increase, but now it is non-linear with increasingly higher amounts of fertilizer leading to smaller and smaller increases in growth rate; in (c) we have the same effect as (b) for small amounts of fertilizer but rather than plateauing with increasingly higher levels of fertilizer, growth rate peaks at intermediate fertilizer levels and actually decreases with increasing fertilizer addition when very high levels of fertilizer are involved. If you are absolutely certain that you have a linear response and you want the most powerful experiment to estimate the gradient of that response then this is achieved by studying only the two extremes of your control variable (0 mg and 30 mg of fertilizer addition in this case). However, study of the extremes gives no information about the shape of the response. If you want information about the shape of the response then you should allocate some experimental units to intermediate values as well as to the extremes.

rapidly growth rate increases with increasing fertilizer (i.e. the slope of the line). In that case, the most powerful design will be to have two extreme treatment groups, one at low fertilizer concentration, and one at high fertilizer concentration, and to split the plants in order to have half (15 plants) in each group. This will give you the best estimate of the slope, and also the highest statistical power to detect a difference between these groups.

However, what happens if you don't know that the relationship is a straight line? Maybe the effect of fertilizer increases to a plateau (as in Figure 6.6b) or increases and then decreases (as in Figure 6.6c). In either of these cases, taking only two treatment groups will give you misleading information about the relationship between growth rate and fertilizer concentration. In the first case, you would not be able to see the important biological (and probably economic) effect that at low levels of fertilizer, a small increase has a big effect on growth rate, while at high levels, the same increase has little effect. In the second case, you might conclude that fertilizer has no effect (or even a negative effect!), even though at intermediate concentrations fertilizer has a big positive effect. The solution is clear: if you don't know the shape of the relationship, and cannot be confident that it is linear, then you need to have intermediate treatment groups to allow you to assess the shape of the relationship. To get the best estimate of the shape of a relationship, you will need to have as many treatment groups as possible, spanning the range that you are interested in, and as evenly spaced as possible.

Even if you are reasonably confident that the relationship is linear, it may be worth hedging your bets to some extent and including one intermediate treatment group. This may slightly reduce your statistical power, but will provide a data set that you can be more confident about.

There are advantages and disadvantages to all methods of selecting levels; choose the design that you think will be best for your particular question and system.

6.4.4 Consider all other ways of increasing power before increasing sample sizes

We have focused on the situation where you have reached the maximum possible sample size and still need additional statistical power. Of course, all of the approaches that we have outlined for increasing the power of a study need not, and should not be reserved for this situation. By considering these possibilities as you plan your study, you may find you can actually use smaller sample sizes than you had initially planned, which is good for you and good for any animals you will no longer need to use. In contrast, if you find that having used all of these approaches your study still lacks power our advice is simple. Don't do it! Find a different study where you can deliver reasonable power.

 There are many ways that the power of a study may be increased without increasing sample sizes.

6.5 Comparing the power of different planned studies

Exactly the same procedures as in the chicken feed example in Box 6.1 can be used to determine the power of any experiment that we can imagine. At its simplest, this may involve comparing designs with different sample sizes, but the same approach can be used to compare different designs, or to examine the effect on power of different strategies for reducing variation. Perhaps you want to know whether the increased cost of using a higher precision cell counter for your assays will be worth the increase in power that it will provide. Or whether the increased variation that is likely to occur if you have several people helping you with your assays of leaf damage will be outweighed by the fact that you will be able to process many more leaves in the time you have available. In principle, to answer these kinds of question all we need to do is decide what we think the world is like, and we can then either simulate experiments with different sample sizes, or different experimental designs. For standard designs it is also possible to obtain power estimates from tables, statistics packages, or websites. Whilst the latter methods are far less time consuming, it is worth noting that for more complex designs doing simulations like those in Section 6.3 may be the only way to calculate power. In Box 6.3 we go through an example of this kind of process. In the Bibliography we provide information on where to go to find out more about power analysis, including some useful websites and more specialist texts.

BOX 6.3 An example of using power analysis to compare different planned studies

When we left our researcher at the animal feed company in Section 6.3, they had determined that the planned study had only about a 50:50 chance of detecting the size of effect that they are interested in. These are not great odds, so they decide to look at the effects of making changes to their design to see whether they can improve the power. The researcher could repeat this whole process using different numbers of imaginary chickens to see what effect changing sample size has on the power of the experiment. Similarly, if it is decided that their original estimate of the effect size is too large they could examine a different effect size, say 250 g, simply by generating their experimental chickens from a distribution with a different mean. Or perhaps they know that by installing an expensive climate control system they could reduce within-sample variation in growth rate by 10%, and they want to know the effect this reduced variation would have on the power of their experiment. Again, the researcher could follow the same procedure, but generate experimental chickens from normal distributions with reduced variation. In principle, this procedure can be used to examine the power of any kind of experimental design, no matter how complex.

In practice, for studies that plan to use standard experimental designs it is generally not necessary to go through this procedure. Instead, there are tables of power values for many of the designs most used by biologists, and many statistics packages will calculate power values too. There are even web-based programs that can estimate power for a given design. All of these methods require estimates of the same things that we needed for our imaginary experiments—the effect size, the amount of variation, the design, and the sample sizes—but are quick and easy to use. Realizing that they are using a standard design our researcher decides that it will be more efficient to look up the power values for a variety of other designs. They examine a range of sample sizes from 5 to 25 chickens in each group, since their research station will only fit 50 birds, and they compare designs with and without the climate control system, to see what effect it might have on power. The results of these investigations are shown in Figure 6.7.

As expected, the power increases with the sample size, and is higher when the amount of variation is reduced. However, increasing sample size above ten birds per group has relatively little effect on power, and the effect of the climate control is not large. Thus the researcher decides they will use ten chickens per group, but will not invest in the expensive climate control system.

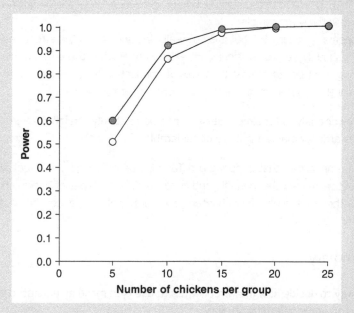

Figure 6.7 The predicted power of experiments to detect an effect of 500 g, with different sample sizes and different amounts of between-individual variation (•, climate control; ○, no climate control). Power increases with sample size, and also is higher for a given sample size when the amount of random variation is smaller. However, the effect of increasing sample size is one of diminishing returns with increases in sample sizes having a large effect on power up to about ten birds per sample, and much less effect above this.

In summary, you can use readily available computer programs, equations in books, or the simulation methods outlined in Section 6.3 to calculate the number of samples that you'll need to collect. But in order to do this, you must know the following:

- How your experiment will be designed.
- Which statistical test you will use on your data.

You must decide the following:

- The effect size that you are interested in.
- The risk you are prepared to run that an effect of this size is really there but your statistics will not detect it (a **type II error** in the jargon).
- The risk you are prepared to run that no effect exists, but your statistics mistakenly suggest that there is an effect (a **type I error**).

Note that you must accept some risk of these errors occurring; there is no certainty in statistics, just acceptably low levels of risk that you've got it wrong. The probabilities of making an error of each type are linked, and, to some extent, under the control of the investigator. In the supplementary material we look at these two types of errors in more detail as well as exploring a more subtle cost of low-powered studies.

In addition, you need to be able to estimate some measure of the variability between individuals in the same sample.

This seems like a long list, but it is not really; in general only the effect size and the variability should give you any significant extra work. Further, this effort in thinking and calculating must be set against the perils of conducting an experiment with too few samples or the cost (and perhaps irresponsibility) of using too many.

> Power analysis for simple designs can be relatively straightforward, and the rewards for mastering it are considerable.

We have emphasized that adopting a different type of design can affect statistical power, but there are other benefits and costs to different types of design and we will overview these in the next four chapters, in order to help you select the best type of design for a given situation.

A **type II error** occurs if there really is an effect of one factor on the variable of interest but your experiment fails to detect it. A **type I error** occurs if there is in fact no effect of a factor, but your experimental results (just by chance) suggest that there is.

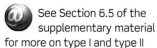 See Section 6.5 of the supplementary material for more on type I and type II errors. Go to: **www.oxfordtextbooks.co.uk/ orc/ruxton4e/.**

■ Summary

- One way to decide how many replicates to use is to make an educated guess based on previous similar studies.
- The alternative is a more formalized method called power analysis. This is not tricky and there are many computer programs that can help you.
- Too few replicates can be a disaster; too many can be a crime. So you need to think carefully about sample sizes.
- Careful consideration of the different sources of random variation in your study can provide ways to increase power without increasing sample sizes.
- Power analysis can also provide a means of deciding amongst different possible experimental designs.

The simplest type of experimental design: completely randomized, single-factor

7

- We begin by introducing one-factor completely randomized designs with two levels of the factor as a simple but powerful means of organizing an experiment (Section 7.1).

- Key to such designs, and to many experiments, is randomization to deal with confounding factors, and we discuss how to do this effectively (Section 7.2).

- We then explore how the number of levels of the factor can often usefully be increased beyond two (Section 7.3).

- We finish this chapter with an overview of the types of situation that would encourage you to adopt this type of design, and the types of situation that might cause you to instead adopt one of the designs described in later chapters (Section 7.4).

So far in this book we have focused on the basic ingredients of good experiments, such as proper replication and effective controls. We now turn to specific experimental designs to see how these ingredients can be combined to answer biological questions. One way to think about different experimental designs is as different ways to deal with all of the potential sources of confounding variation and noise. As we will see (in Section 7.4), the design that you choose will be determined by the question that you are asking, and the biology of your study system (in particular the potential sources of variation); and will determine the kind of statistical analysis that you are able to do. At first sight, the experimental design literature can seem daunting, and full of complex terms and jargon. However, there is really nothing difficult about the ideas behind these terms. We will introduce and define the commonest terms as we go along. In this chapter we will go through the process of setting up the simplest experimental design, bringing together ideas from previous chapters. In subsequent chapters we will build on this simple design, at each stage focusing on how the additional complexity of the design deals with variation and affects the questions that we can ask.

7.1 Completely randomized single-factor designs

Let's start with a simple biological question:

Does plant feed affect the growth of tomato plants?

To answer this question we are obviously going to need to apply some plant feed to some tomato plants and measure some aspect of their growth, perhaps their dry weight after a few weeks of growing. We are also going to need some form of control so that we know what growth we might expect in the absence of the plant feed. In this case our control would be tomato plants that are not given the plant feed but are otherwise handled in an identical way.

Now we need to introduce some jargon. Our experimental manipulation affects a single factor, the presence or absence of plant feed, and so this design is known as a **one-factor design** (you will also see it referred to as a **one-way design**). Our factor has two **levels**, since our experiment has two treatments ('plant feed' and 'control').

Of course for all the reasons outlined in Chapter 4, we can't just compare the growth of two plants, one with and one without plant feed. We need several plants in both our 'plant feed' and 'control' treatment groups, or to use more jargon, our experiment needs to be replicated. In this case we have 80 tomato plant seedlings all ready to be used in our experiment. All we need to do is allocate them to our two treatment groups, apply the plant feed to those in the plant-feed treatment group, and our experiment is ready to go. However, an important question remains, how should we decide which plants go into which treatment group? In this simple experimental design, the answer is equally simple: **randomization**.

> A simple one-way design allows us to compare the effects of different levels of a single factor.

7.2 Randomization

7.2.1 Randomizing study subjects

The experimental subjects of any study will vary for many reasons that are nothing to do with the factors that we are interested in. The chickens in Chapter 4 produced eggs with different shell thickness for many reasons that were nothing to do with the kind of feed they had been given. Similarly, in our current study, plants will vary in their growth for all sorts of reasons that have nothing to do with whether they are treated with plant feed or not. Our challenge is to separate the effects we are interested in (in this case any effect of plant feed) from all of this background noise. We want to ensure that there are no systematic differences between the plants in our two treatment groups from any of these other sources of variation in growth, otherwise any differences caused by our experimental treatment could be confounded with these other factors and cause us to draw erroneous conclusions. Imagine that there are small

An **n-factor** or **n-way design** varies n different independent factors and measures responses to these manipulations. For example, an experiment looking at the effects of both diet and exercise regime on dogs' health would be a two-factor design. If we consider three different diets, then there are three **levels** of the experimental factor 'diet'. If n is greater than one (i.e. multiple factors are involved), then the experiment can be referred to as a *factorial* experiment.

Randomizing the allocation of experimental subjects to experimental groups simply means that each subject is allocated at random to a group, with any subject being equally likely to end up being allocated to any of the groups. This process is also called *random allocation, simple random allocation, random assignment,* or *random placement.*

 Q 7.1 We have assumed that the experiment should have a concurrent control (in the jargon we introduced in Section 2.3.1). Would you be tempted to halve the size of the experiment and use a historical control instead?

differences in the quality of the compost in each pot that our plants are growing in, and that this has an effect on the growth of the plant. If, for whatever reason, the plants with slightly better compost end up in the plant-feed group, whilst those with slightly worse compost end up in the control group; then we are likely to see a difference in growth rate between our groups, even if the plant feed has no effect at all. This is true for any of the other factors that might affect plant growth. So our task is to ensure that the only systematic difference between the plants in each treatment group is the factor of interest (presence or absence of plant feed). The simplest solution to this problem is to let chance decide: to randomly allocate subjects to each treatment group (say by tossing a coin for each plant). By randomly allocating plants to treatment groups we randomly spread the variation due to other potentially confounding factors across treatments. This will minimize the chance of any systematic differences in other factors affecting our conclusions. The major advantage of randomization is that it not only deals with sources of variation that you might be suspicious about, such as compost quality, but also deals with all of the many other potentially confounding factors that you don't know about: everything gets randomized. It will probably not surprise you to learn that this use of random allocation to treatments is what leads to this type of study being referred to as a *completely randomized design*, since subjects are assigned to treatment groups completely at random.

 Randomization is the key to avoiding confounding factors in planned experiments.

7.2.2 Randomizing other aspects of your study

Randomizing the allocation of experimental subjects to treatments is a simple and powerful way of avoiding systematic differences amongst treatment groups. However, it is only the start. Our tomato plant study is going to take place in a greenhouse. Each plant will be placed in an individual pot in the greenhouse. The size of the experiment means that our plants will need to be spread over two benches within the greenhouse. After 2 months, we will harvest the plants and measure their dry mass. There are many aspects of this experimental set up that may affect plant growth. Maybe the greenhouse is slightly warmer or lighter at one end than the other, or maybe the glass above one of the benches is slightly older and so lets through a different colour of light than the glass above the other bench. The list could obviously go on. If we put all of our plant-food-treated plants on one bench, and the control plants on the other bench, then our replicates have effectively become pseudoreplicates, since treatment effects are confounded with systematic differences between benches. The easiest way to avoid this is to carry out further randomization. We can randomize the allocation of plants to benches, and the position of plants on a single bench, such that any plant has an equal probability of being in any part of the greenhouse.

However, the need to randomize does not just apply to the setting up of the experiment. It can equally be applied to taking the measurements at the end of a study. There are numerous reasons that can lead to the accuracy of measurements differing through time. Maybe the spectrophotometer that you are using is old, and gets less accurate as time goes on. Or maybe you have spent 10 hours looking down a microscope counting parasites, and the inevitable tiredness means that the later counts are less accurate. Or maybe after watching 50 hours of great tit courtship behaviour on video you become better at observing than you were at the beginning. In the current study, perhaps the balance that we are using to weigh our dry tomato-plant samples becomes less accurate as time goes on, or the dried samples take in moisture through time, slowly increasing in mass. Whatever the reason, this means that if you take all the measurements on subjects from one treatment group first and then all those from another treatment group, you risk introducing systematic differences between the groups because of the changes in the accuracy of your measurement methods. It is far better to organize your sampling procedure so that subjects are measured in a random order (see Section 11.1 for more about this sort of problem). In short, proper randomization at all stages of study is one of the most simple and powerful tools available to us to avoid problems of systematic bias and confounding variables. It ensures that sources of non-independence between subjects do not creep into our study and lead us to unwittingly pseudoreplicate. We emphasize the word 'proper' here, because inadequate randomization is probably one of the most common flaws in experimental design, in experiments by everyone from undergraduates to the most eminent professors.

> Randomization isn't just the way to avoid confounding variables when allocating experimental subjects to different treatments; it is the key to avoiding confounding variables creeping in at all stages of your experiment.

7.2.3 Haphazard allocation

The major problem that arises in randomization is that, for many people, when they say that they randomly allocated subjects to treatments what they actually mean is that they employed *haphazard* allocation. So what is the difference? Let's briefly consider a different study. Imagine we have a tank full of 40 hermit crabs that we want to use in a behavioural experiment. The experiment requires that we allocate them to one of four different treatment groups. Random allocation would involve something like the following:

- Each crab would be given a number from 1 to 40.
- Pieces of paper with numbers 1 to 40 are then placed in a hat.
- Ten numbers are drawn blindly, and the crabs with these numbers allocated to treatment A.
- Ten more are drawn and allocated to treatment B, and so on until all crabs have been allocated.

Of course, we could equally have drawn the first number and put that crab in treatment A, the second in B, third in C, fourth in D, and then repeated until all treatments were filled. However, what is important is that each crab has the same chance as any other of ending up in any treatment group and so all of the random variation between crabs is spread across treatments. Another way to think about a random sample is that the treatment group selected for one subject has no effect whatsoever on the treatment group selected for the next subject.

This randomization procedure contrasts with a haphazard sampling procedure. A typical haphazard sampling procedure would involve placing one's hand in the tank and grabbing a crab without consciously aiming for a particular individual. *This will not give you a random sample.* Even if you shut your eyes and think about your bank balance, this is still not going to give you a random sample. The reason for this is that there are a large number of reasons that could cause the first crabs to be picked out to be systematically different from the last crabs. Perhaps smaller crabs are better at avoiding your grasp than larger ones. 'Hold on,' you say, 'what about if I pick the crabs out in this way and then allocate them to a group without thinking, surely this will give me random groups?'. Well, maybe it will, but probably it won't, depending on how good you are at not thinking. It is very tempting to subconsciously think, 'I've just allocated a crab to treatment *A*, so I guess the next one should be given a different treatment'. This is not random. So if you really want random groups, the only way to get them is to randomize properly, by pulling numbers from a hat, or generating random sequences on a computer. It may seem like a nuisance, and it may take you an extra half an hour, but an extra half hour is a small price to pay for being confident that the results that you obtain after weeks of work actually mean something.

 Always randomize properly, don't be tempted into using easier procedures which are open to bias.

7.2.4 Balanced and unbalanced allocation

Now that we understand the importance of randomization, we allocate our tomato plants at random to our two different treatment groups. One way to do this would be to take each plant in turn and flip a coin. We then allocate the plant to the plant-feed group if we see heads, and the control is we see tails. This procedure would certainly achieve a random allocation of plants, but because of the vagaries of random sampling we are very likely to end up with slightly different numbers of plants in each treatment group: perhaps 36 in one and 44 in the other. In statistical jargon, our experiment is **unbalanced**. This is not a fatal flaw in our experiment, but it does have drawbacks. Generally the statistical methods that we will ultimately use to analyse the data are most powerful, and least influenced by violations of assumptions, when there are equal numbers in each group. Only in unusual circumstances should we deliberately seek to use unequal numbers. Thus aiming for equal numbers, or as a statistician would say for a **balanced design**, should be a general tenet of your experimental designs. A better way to assign the plants would be to add a constraint that the final number in each group must be 40. Practically, this could be done by numbering all of the plants, then putting these numbers on identical pieces of card in a bag, mixing them thoroughly

A **balanced** experimental design has equal numbers of experimental units in each treatment group; an **unbalanced** design does not.

You can find one example of when you might use unbalanced groups for ethical reasons in Section 7.2.4 of the supplementary material. Go to: **www.oxfordtextbooks.co.uk/ orc/ruxton4e/**.

then drawing them out. The first 40 numbers to be drawn would be assigned to the first treatment group, and so on.

We have now designed our study, so let's revisit some of the jargon. We have used *a completely randomized one-factor design, with a factor that has two levels. Our design is also fully replicated and balanced.* You can see how the terminology can become complicated very quickly, but the underlying logic is straightforward. This design allows us to look for differences between two groups. As we emphasized in Statistics Box 2.1, the final component we need to think about as we plan our study is how we will analyse our data. A major advantage in using simple, well understood designs is that they lead very naturally to statistical tests that we can use to answer the question once we have our data. The most obvious choices to analyse data from a single-factor design with two levels of the factor would be to use a *t*-test, or its non-parametric alternative, the Mann-Whitney U test.

➜ Always aim to balance your experiments, unless you have a very good reason not to.

7.3 **Factors with more than one level**

Now that we understand this simple design we can start to expand it to address other questions. One obvious way we might expand our study is to add additional levels for our factor. Thus, suppose that we wanted to explore the effects of supplying the plant feed at different rates (let's call the rates *low*, *medium*, and *high*). In principle we could repeat the study above for each rate in turn, but a much better alternative would be to examine the feeding rates in the same study, by including four groups: (i) control (no plant feed), (ii) feed provided at low rate, (iii) feed provided at medium rate, and (iv) feed provided at high rate. This is still a one-factor design because we are still only manipulating a single factor (rate of plant feed delivery), but now our factor has four levels instead of two. There are multiple advantages to this; most obviously it allows us to compare each feeding rate to the others, as well as to the control. That is, we can answer several questions with a single study:

Q 7.2 If we carried out three separate experiments, one immediately after the other, each comparing a control group to either a low, medium, or high rate of plant-feed application, couldn't we still compare the effects of different feeding rates?

Does providing plant feed at a low rate affect the growth of tomato plants?

Does providing plant feed at a medium rate affect the growth of tomato plants?

Does providing plant feed at a high rate affect the growth of tomato plants?

Do the three feeding rates differ in their effect on the growth of tomato plants?

This single four-level experiment is also a more efficient use of plants than three two-level ones, since we can compare the three feeding rate treatments to the same control plants rather than needing three separate sets of control plants. Of course, everything that we discussed above about setting up the simple design applies equally to this design. We would randomly allocate our plants to the treatment groups, and randomize everything else that we could. We would also probably have equal numbers of plants in each treatment group, ensuring that our design is balanced. Putting all of this together we could describe our experiment as a *completely randomized one factor design with four levels of the factor and that it is fully replicated and balanced*. As for the analysis, in this case your design naturally leads to either a one-way analysis of variance (ANOVA) or a Kruskal–Wallis test.

 A larger experiment with several levels of the same factor is always more efficient than a collection of smaller experiments.

7.4 Advantages and disadvantages of complete randomization

Imagine that you have twenty plants for a growth trial, and you split them randomly into four groups, each of which is given a different feed treatment. If all goes well, you will have twenty growth rates to analyse at the end of the experiment. However, if a plant becomes diseased, or is knocked off the bench, or some other catastrophe befalls it such that the growth rate of that plant cannot be obtained or no longer provides a fair reflection of the treatment, then you will have fewer measurements to work with. The lost or omitted cases are generally called *missing values* or *drop-out cases*. Obviously, experiments should be designed to minimize the likelihood of such drop-outs occurring. However, accidents do happen, so you should seek experimental designs where a small number of drop-outs do not have a devastating effect on the usefulness of the remaining data.

The attraction of complete randomization is that it is very simple to design. It also has the attraction that the statistics that will eventually be used on the data are simple and robust to differences in sample sizes, so having missing values is less of a problem. Also, in contrast to some designs that we will see in Chapter 10, each experimental unit undergoes only one manipulation; this means that the experiment can be performed quickly and that ethical drawbacks to multiple procedures or long-term confinement are minimized. It tends to be that the probability that a subject will drop out of the study increases with the length of time over which they need to be involved in the study, so the drop-out rate should be lower than in some other types of experiments.

The big drawback to full randomization is that we are comparing between subjects. As we saw in previous chapters, between-subject variation due to random factors makes it difficult for us to detect the effects of the manipulations that we have carried out, and so reduces our statistical power. If the growth rates of tomato plants are highly variable due to other factors, this may make it very hard for us to detect the effects of our different feeding regimes. With a completely randomized design, then the only real option to deal with this problem is to increase sample size, with all of the associated ethical, conservation, or financial costs this may incur (and the experiment will also use more of our time). An alternative approach is to consider using a more complex experimental design that explicitly takes some of these sources of variation into account. We will explore these in Chapters 9 and 10; but before that we will spend Chapter 8 exploring why you might often want to vary more than a single factor within an experiment, and how you might most effectively design experiments that incorporate multiple factors.

Q 7.3 What could we do in our tomato plant trial to minimize drop-outs?

Q 7.4 A researcher is interested in whether competition affects the development time of beetle larvae that complete their larval development within mung beans. To do this she allows adult females to lay eggs on mung beans for several hours, and then examines each bean, counting the number of eggs that it carries. She randomly picks 50 of the beans carrying single eggs for her low-competition treatment, and 50 of the beans carrying two eggs for her competition treatment, and then measures the time larvae take to develop in these two treatments. Do you see any problems in this approach? How might it be improved upon?

➡️ If between-subject variation is low, then complete randomization can be quite powerful, hence it is commonly used in laboratory studies. It is also attractive if you expect high drop-out rates among subjects. Field studies and clinical trials tend to suffer more from between-subject variation, and so for these the designs discussed in Chapters 9 and 10 may be more attractive.

Summary

- A simple one-way design allows us to compare the effects of different levels of a single factor.
- Randomization is the key to avoiding confounding factors in planned experiments.
- Randomization isn't just the way to avoid confounding variables when allocating experimental subjects to different treatments; it is the key to avoiding confounding variables creeping in at all stages of your experiment.
- Always randomize properly, don't be tempted into using easier procedures which are open to bias.
- Aim to have a balanced (or near-balanced) experiment with the same (or very similar) numbers of experimental subjects in each treatment group.
- A larger experiment with several levels of the same factor is always more efficient than a collection of smaller experiments.
- If between-subject variation is low, or drop-out rates are expected to be high, then complete randomization is an attractive technique. Otherwise the methods considered in Chapters 9 and 10 may be more attractive.

Experiments with several factors (factorial designs)

8

- In this chapter we firstly introduce the concept and terminology associated with extension of the randomized design of Chapter 7 to include multiple factors (Section 8.1).
- We then discuss the concept of interaction between those factors (Section 8.2).
- We then add a note of caution against confusing levels and factors (Section 8.3).
- We introduce split-plot designs and Latin square designs and explain their attractions and limitations (Sections 8.4 and 8.5, respectively).
- We finish the chapter by discussing how the design concepts raised in this chapter link to statistical analysis (Section 8.6).

8.1 Randomized designs with more than one factor

Let's continue to think about the tomato plant project from Chapter 7. Imagine that (as well as the effects of plant feed) we are also interested in whether the application of insecticide has any effects on tomato plant growth. How should you go about exploring the effects of insecticide? The first option is to simply do what you did with plant feed and carry out a second experiment, again using a one-factor design with the presence or absence of insecticide as our experimental factor. Such an experiment would address the following question:

Does the application of insecticide affect the growth of tomato plants?

Whilst there is nothing wrong with this in principle, there is another possibility, which is to carry out a single experiment that looks at the effects of both plant feed and insecticide at the same time. Since we are exploring the effects of more than one factor, this fits the definition of a factorial experiment given in Chapter 7. How should we set up such an experiment? First we would allocate each of our tomato plant seedlings at random to one of the four plant feed treatments ('control', 'low', 'medium', or 'high'). This is exactly what we did in Chapter 7. However, we would then take the seedlings allocated to one of the feed rates, say the control treatment, and randomly allocate them to one of the two insecticide treatment groups ('insecticide applied' or 'no insecticide applied'). We would repeat this for the other three plant feed treatments. Now we are varying two factors, the rate of plant feed application and whether or not we apply insecticide. This sort of design is called a *two-factor design* (or *two-way design*). The way that we have set it up means that it is also a *fully cross-factored* (or **fully crossed**) *design*. This means that we have all possible combinations of the two factors; or to put it another way, we have plants with and plants without insecticide in all of our plant feed treatment groups. In contrast, if for whatever reason we could only apply insecticide to plants in two of the plant feed treatment groups, but had plants without insecticide in all four plant feed treatment groups, then our design would have been described as an *incomplete design*. The statistical analysis of fully cross-factored designs is easy to do—in this case we would probably choose a two-way analysis of variance (ANOVA). In contrast, analysing incomplete designs is significantly more difficult. Avoid incomplete designs whenever possible. If you need to use an incomplete design for practical reasons (such as a shortage of samples) then get advice from a statistician first.

A **fully crossed** design means that all possible combinations of treatments of the factors are implemented. Thus if in a three-factor experiment, factor *A* has two levels, factor *B* has three levels, and factor *C* has five levels, then a fully crossed design will involve 30 different treatment groups (2 × 3 × 5).

 For factorial experiments, your analysis will be a lot easier if you have a complete design.

8.2 Interactions

We have arrived at a fully cross-factored, two-factor design, with four levels for the first factor (plant feed), and two for the second (insecticide). This will give us eight treatment groups in total (4 × 2), and, as long as we have the same number of plants in each treatment, and more than one plant in each treatment too, our design will be balanced and replicated. You can easily imagine how you can get three-factor, four-factor, and even more complicated designs. In general, it is best just to imagine these; for experiments that you actually have to do, avoid making them so complex that you get overwhelmed either collecting the data or trying to analyse it.

If simplicity is the key to good design, then you might ask, 'Why do a two-way factorial design in the previous case; why not perform the two separate one-factor designs?' The main attraction of using a two-way design is that it allows us to answer multiple questions within the same experiment. In this case the main questions would be:

Do the plant feed treatments affect the growth of tomato plants?

Do the plant feed rates differ in their effects on the growth of tomato plants?

Does the application of insecticide affect the growth of tomato plants?

Does any effect of insecticide on growth depend on the rate at which the plants are being given plant feed?

The first three questions are about what statisticians refer to as the **main effects** of each factor (the overall effect of the plant feed and the overall effect of insecticide). The final question is about the **interaction** between the factors, what statisticians call the *interaction effect*, and this can only be answered with a factorial design.

In many biological studies, the most interesting questions are specifically about the interactions between factors rather than their main effects. For example, we might hypothesize that older people respond more slowly to a given drug therapy than younger people. If we put this into statistical terms, we are suggesting that the effect of one factor will depend on the level of another; there is an interaction between the two factors. The important thing to stress here is that, if your hypothesis is about the interaction, you must test for that interaction. And if you need to test the interaction, you need to use a design that allows you to test the interaction. This sounds obvious but it is easy to go wrong if you are not careful; we explore these issues in more detail in Box 8.1 and Statistics Box 8.1.

So our more complex two-factor design has allowed us several questions, including questions about interactions between factors that would not be possible with the simpler one-factor design. It has also saved us from some difficult decisions. Thus, suppose we had already completed our single-factor plant-feed study from Chapter 7 and are now planning a separate experiment to look at insecticide. An obvious question is should we be using plants that we also give plant feed, and if so, at what rate should they be fed? By carrying out a single two-factor experiment we avoid this decision, as we are looking at the effect of insecticide under all four plant-food conditions at once. Indeed, this also means that if we do see an overall effect of insecticide in our study we can be more confident about how general any effect of insecticide is likely to be, since we are testing it under four different plant feed conditions (the same can obviously be said for effects of plant feeding, since they are being examined with and without insecticide). This issue is discussed more extensively in Box 8.1. And there is one final advantage: our two-factor study has almost as much power to test for the effects of plant feed as a one-factor study with the same total number of plants. It also has almost as much power to test for insecticide effects as a one-factor study of insecticide with the same total number of plants. Thus, our two-factor study can examine both the effect of plant feed and the effect of insecticide using about half the number of tomato plants as would be required if we were to carry out separate experiments for each factor.

On hearing this argument, you are probably tempted to add more and more and more factors to your study. We would caution against this: if you vary too many factors simultaneously, your experiment will necessarily be very large, requiring a large number of subjects and a large amount of effort. Also, interpreting interactions between more than two factors can be challenging.

Let us imagine that we want to investigate whether dependent factor A is influenced by independent factors B and C. If the effect of B on A is affected by the value of C, or equivalently if the effect of C on A is affected by the value of B then there is an **interaction** between the factors B and C. Alternatively if the value of A is affected by the value of B in the same way regardless of the value of C, then we can say that there is a **main effect** due to factor B. Similarly, if A is affected in the same way by the value of C regardless of the value of B, then there is a main effect due to C. If there is an interaction, then by definition there are no main effects. If there is no interaction then both, one, or no main effects might be present.

Q 8.1 Let's illustrate the problem of interpreting an interaction involving three factors. Imagine that in the tomato plant experiment we explore the effects of tomato plant variety (two levels), food applied (two levels: whether a plant food was added to the growing medium at low or high dose), and insecticide treatment (two levels: whether an insecticide was sprayed on the plants or not) on growth rate. Our subsequent statistical analysis informs us that there an interaction between all three factors; can you describe biologically what this means?

BOX 8.1 Interactions and main effects

In the main text we claimed that one advantage of carrying out experiments with two or more factors was that this would allow you to look at both the *main effects* of the factors and also their *interactions*. Understanding what these actually are is fundamental to understanding these more complex designs. However, these concepts are also some of the most misunderstood statistical ideas. In this box we will explore these concepts in a little more detail in the hope of showing that they are both easy to understand and extremely useful.

Let's go back to thinking about the effects of feeding treatments on growth rate in different varieties of tomato plants. Imagine that we want to investigate two feeding regimes, high and low food, and we want to compare two varieties of tomato, which we will call varieties *A* and *B*. This will give us four different treatments (both varieties with both feeding regimes). Figure 8.1 shows several of the possible outcomes of an experiment like this.

Let's begin with the situation in Figure 8.1a. Here we can see that for both varieties of tomato, the high-feed treatment leads to faster growth than the low-feed treatment. However, if we compare the two varieties of tomato when they are fed identically, we see no difference. Statistically, we would say that there is a significant main effect due to feed treatment, but no main effect due to tomato variety. Biologically, we would say that growth rate is affected by how much you feed the tomatoes, but not by particular tomato variety.

The situation in Figure 8.1b is a little more complicated. Here again, we see that for both varieties the high-feed treatment leads to faster growth than the low-feed treatment. Although now, if we compare the two varieties under the same feed treatment, we find that variety *A* grows slower than *B*. Statistically, we would now say that we have significant main effects due to both feed treatment and tomato variety; biologically, we would say that tomato growth rate is affected by how much you feed them and also by the tomato variety.

Figure 8.1c looks very similar, but differs in one very important respect. Again, we can see that for both varieties, high feeding increases growth rate. Similarly, if we look at the growth rate of the two varieties when they are given lots of food we see that, as in the previous case, variety *A* grows slower than *B*. However, if we compare the varieties at the low-feed level, we see no difference between them. What does this mean? This is what statisticians refer to as an interaction. Biologically, it means that the effect of the variety of tomato on growth rate is different under different feeding regimes, or to be a bit more technical, the effect of one factor (variety) depends on the level of the other factor (feeding level). In our example, the difference is quite extreme; there is no difference between the varieties at one food level, but a clear difference at another.

In Figure 8.1d we see a less extreme case. We do see a difference between the varieties under both high- and low-food regimes, but under low food the difference between the varieties is much smaller. Statistically, this is still an interaction—the effect of one factor still depends on the level of the other—or to put it

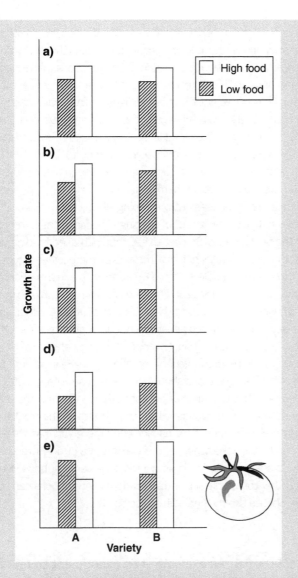

Figure 8.1 The growth rates of tomato plants in a two-factor experiment. One factor is plant variety (with two levels: *A* and *B*); the other factor is feeding regime (two levels: *high* and *low*). In (a) there is a main effect of feeding regime only (that is, there is no difference in growth rate between the two varieties when fed at the same level). In (b) both feeding regime and variety have separate main effects on growth rate. In (c), (d), and (e) there is an interaction between the two factors, such that the effect of one is different for different levels of the other.

biologically, the effect of variety on growth rates depends on which food treatment we consider.

We might even see a third type of interaction, as shown in Figure 8.1e. Here the growth rate of variety *A* is slower than *B* when they are given lots of food, but faster than *B* when they are given only a little food, so not only does the size of the difference in growth rates depend on the food treatment, but so does the direction of the effect.

So why should we care about interactions? One reason is that a significant interaction can make it very difficult to interpret (or to even talk about) the main effect of a treatment. Imagine we got the result in Figure 8.1e. What is the main effect of food level? In one variety it is positive, and the other negative, so it is meaningless to talk about the main effect of food level. It only makes sense to talk about the effect of food if we state which variety we are referring to. Similarly, in Figure 8.1d if we want to talk about the difference between the varieties we need to state which feeding regime we are comparing them under. We can't really talk about an overall difference between the varieties in any meaningful way. Even in Figure 8.1b it is hard to say exactly what the main effect of feeding is because the size of the effect is different for the two varieties. So we can say that increased feeding has an overall positive effect on growth rate, but that the size of the positive effect depends on the variety. Only in situations like in Figure 8.1a can we make a confident statement, such as the overall effect of increased feeding is to increase growth rate by a specified number of units. Another way to think about this is that our factorial design has allowed us to say more about the generality of our treatment effect. If we had only carried out the experiment on a single variety, then any effect we detected might be very specific to the particular variety that we chose. By including a second variety, we can start to ask questions about how general any effects of the treatment are. If the treatment has the same effect for both varieties (i.e. there is no interaction), then we can be more confident that the effects of the treatment are likely to be more general. Of course, if we had included more varieties then we could be even more confident. On the other hand, if we detect a strong interaction between treatment and variety, then it tells us we must be very cautious in extrapolating our results to other tomato strains. Interpretation of interactions can sometimes be tricky, and we discuss this further in Statistics box 8.1.

STATISTICS BOX 8.1 An example of an experiment to test for an interaction

Suppose that you are interested in sexual selection in sticklebacks. You might hypothesize that because males invest heavily in sexual displays and competing for females, they may be less good than females at coping with parasitic infections. One possible prediction of this is that any deleterious effects of parasites will be greater on male sticklebacks than on females. This is a hypothesis about an interaction. We are saying that we think that the effect of parasitism will depend on the sex of the fish. In other words the effect of the first factor (parasitism) will depend on the level of the second (sex). By now, you will immediately see that one way to test this hypothesis would be to have a two-factor design, with infection status (parasites or no parasites) as one factor, and sex (male or female) as the other.

You might then measure the weight change in fish during the experiment as an estimate of the cost of parasitism. As an aside, you might be wondering why you need the fish with no infections—wouldn't it be enough to simply have infected males and females and compare their change in weight in a one-factor design? The problem with this is that we would have no control. If we found that fish changed in weight, we wouldn't know whether this was due to the parasites or something else (for example, the feeding regime). So, to measure the effect of parasitism, we need to compare fish with and without parasites. Now if we performed the experiment this way, it would be a relatively simple matter to test for the interaction between sex and infections status, and if the interaction was significant, this would support our prediction (but remember that a significant interaction could also arise if parasites had a bigger effect on females than males, so it is essential to look at plots of the data to make sure that the trend is the way round that you predicted).

Now this (hopefully) seems straightforward to you, but we claimed earlier that people often make mistakes, so what do they do wrong? The most common mistake is to treat this experiment as if it were two single-factor experiments done on the males and females separately. The researcher might analyse the difference between males with and without parasites, and find that males with parasites lose significantly more weight. The analysis could then be repeated on the females, and may perhaps find no significant difference between the weights of the two types of female. From this the researcher would then conclude that there is a difference in the effect of parasites on males and females. This seems logical, but it is wrong. The reason it is wrong is that the researcher has only tested for the main effects of parasitism, but not the interaction between parasitism and sex. By doing separate tests, the researcher has looked at the difference in male fish caused by parasites, and at the difference in female fish caused by parasites, but has not directly compared the difference in male fish to the difference in female fish. You may be thinking that we are being needlessly fussy, but in Figure 8.2 we show how this can make a difference. Figure 8.2 shows the results of an experiment like the one that we have just described. You can see that the weight loss of male fish is greater when they are parasitized than when they are not, and a statistical comparison of these means would show a significant difference between them. In contrast, for the females, the statistical test says there is no significant difference between the parasitized and unparasitized individuals. However, if we look at the figure, we can see that mean values for the females are exactly the same as those for the males. What is going on? Why is the male difference statistically significant, but the female difference not? Well there are several possible reasons, but one is that maybe the sample sizes for males and females are different. Suppose we had only half as many females as males, this would mean our statistical power to detect a difference between the female fish would be much lower than for the male fish, leading to us observe a significant difference in male fish but not in female fish. However, this is hardly good evidence for concluding that there really is a difference in the effects of parasites on males and females. Looking at the graphs, common sense

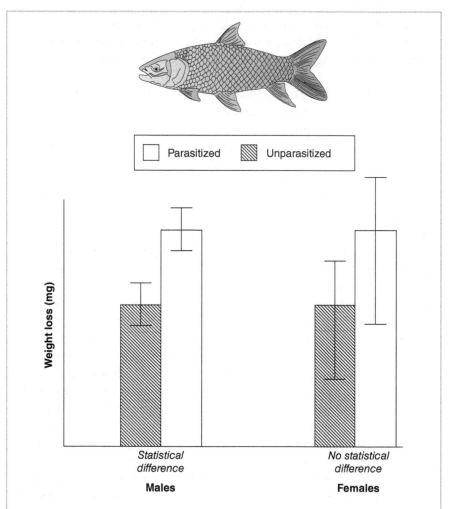

Figure 8.2 The mean and standard error of weight loss in fish, segregated by sex and parasitism status. We see that for both male and female fish the mean level of weight loss is greater for parasitized fish than unparasitized ones. Indeed, the mean weight loss for parasitized fish is the same for both sexes. This is also true for unparasitized fish. What is different between the sexes is the standard error (a measure of the effect of within-sample variation on our confidence that our sample mean is a good estimate of the population mean). For both parasitized and unparasitized fish, variation in weight loss seems much more important in females than males. The power of a statistical test is reduced when the effect of such variation is higher, so a test might find a significant effect of parasitism status on weight loss in males but not in females. However, as plotting the means makes clear, this difference between two separate tests should not automatically be interpreted as suggestive that parasitism status and sex interact in their effect on weight loss.

tells us that there is absolutely no good evidence for a difference in the way males and females respond to parasitism and if the researcher had tested the interaction directly using an appropriate statistical test, this is the conclusion that they would have drawn. This is not just a problem caused by unequal sample sizes; it can arise for any number of reasons causing greater intrinsic variation in some groups than others. In the end the only way of being confident that you have (or don't have) an interaction is to test explicitly for that interaction.

Notice in this example that sex is not a randomly allocated factor (nor was variety in our tomato plant example). That is, individuals are either intrinsically male or female; we do not randomly allocate subjects their level of the factor 'sex'. Despite this, we can still carry out the experiment and perform the analysis in exactly the same way as if the factor were randomly allocated. The only difference comes when we interpret our results. Because we did not allocate sex randomly, we have to be on our guard against confounding factors. If we find an interaction or main effect involving sex, we need to be on our guard that the driving factor might not be sex itself but something that correlates with sex but which we have not included in our statistical model. For example, it might be that it is not sex that has a strong effect but body mass, and males tend to have higher body mass than females.

Q 8.2 An eminent professor hypothesized that plants from old tin mine sites will have evolved increased ability to deal with tin poisoning. In a laboratory study, plants collected from mine sites grew bigger than plants from non-mine sites when grown in soil containing high tin levels. From this, the professor concluded that his hypothesis was correct. A bright young research student questioned this conclusion on the grounds that there was no evidence the increased growth rate had anything to do with tin; maybe the mine plants had evolved to grow bigger for some other reason. The professor repeated his experiment using tin-free soil and found no difference in the size of plants from the two groups. The professor now concluded that his hypothesis was certainly correct. Do you agree?

➡ Often hypotheses we are interested in testing involve the interaction between factors. If you are interested in such a question, then you must ensure that your experimental design and statistical analysis allow you to directly explore this interaction.

8.3 Confusing levels and factors

One confusion that can arise with complicated designs is the distinction between different factors and different levels of a factor. You need to avoid such confusion in order to analyse and interpret a study correctly. In the tomato example, it is easy to see that the different feeding rates are different levels of a single factor. In contrast, what if we had performed an experiment with five different brands of plant feed? Now it is less clear; are the plant feeds different factors or different levels of the same factor? If we had performed this experiment and had a treatment for each plant feed, then we must think of the five plant feed treatments as five levels of a factor called plant feed type, and we would have a one-factor design. But suppose we then carried out an experiment with only two of the plant feeds and set up the following four treatments: no plant feed, plant feed A only, plant feed B only, plant feed A and plant feed B together. Now it is better to view the plant feeds (A and B) as two separate factors with two levels (presence or absence), and analyse the experiment as a two-factor design. In Figure 8.3, we show a number of possible designs with their descriptions to help you get used to this terminology.

 Take care to avoid confusing levels and factors.

1.

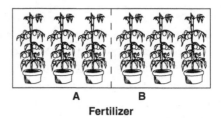

A   B
Fertilizer

Replicated 1-factor design with 2 levels of the factor
(fertilizer type).
This can answer the question:
a) Do the fertilizers differ in their effect?

2.

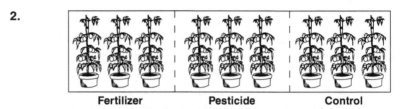

Fertilizer   Pesticide   Control

1-factor design with 3 levels of the factor (type of cultivation).
This can answer the questions:
a) Does fertilizer affect plant growth?
b) Does pesticide affect plant growth?
c) Do fertilizer and pesticide differ in their effect on plant growth?

3.

Pesticide

No
Pesticide

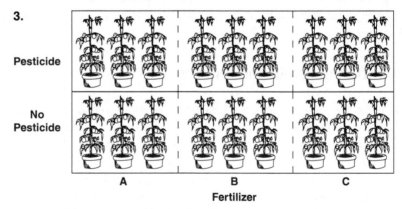

A   B   C
Fertilizer

2-factor design with 3 levels of the 1st factor (fertilizer type) and 2 of
the 2nd factor (pesticide use).
This can answer the questions:
a) Do the fertilizers differ in their effect on plant growth?
b) Does pesticide affect growth rate?
c) Does the effect of pesticides depend on the type of fertilizer?

Figure 8.3 Different experimental designs allow different questions about
the system under study to be explored. In this case, we have three different
experimental designs that allow exploration of the influence or influences on
growth rate of tomato plants. The more complex the design, the more complex the
range of questions it can be used to address.

8.4 Split-plot designs (sometimes called split-unit designs)

The term **split plot** comes from agricultural research, so let's pick an appropriate example. You want to study the effects of two factors (the way the field was ploughed before planting and the way pesticide was applied after planting) on cabbage growth. Three levels of each factor are to be studied. We have at our disposal six square fields of similar size.

The full randomization method of conducting this experiment would be to divide up each field into, say, six equal sections then randomly allocate four of the 36 sections to each of the nine combinations of ploughing and pesticide application (see Figure 8.4 for a picture of what the experiments would look like). The split-plot method would be to allocate two fields at random to each ploughing type, and then plough entire fields the same. We would then take each field and randomly allocate two of the six parts of that field to each pesticide treatment. Surely, this is a worse design than full randomization—why would anyone do this?

The answer is convenience, and sometimes this convenience can be bought at little cost (although the statistical analysis of split-plot designs is more involved). In practice it is difficult to organize for different parts of a field to be ploughed in different ways; our split-plot approach avoids this nuisance. In contrast, applying pesticides, which is done by a person rather than a machine, can easily be implemented in small areas. The price we pay for adopting a split-plot design is that it is much better able to detect differences due to pesticide use (the **sub-plot factor**), and the interaction between herbicide and ploughing, than it is to detect differences due to ploughing (the **main-plot factor**). This may be acceptable if we expect that the effects of ploughing will be much stronger (and so easier to detect) than those of pesticide application, or if we know that ploughing will have an effect but what we are really interested in is the interaction between ploughing and pesticide use.

Split-plot designs are not restricted to agricultural field trials. Imagine exploring the effects of temperature and growth medium on yeast growth rate. The experimental units will be Petri dishes inside constant-temperature incubators. It would be natural to use temperature as the main-plot factor in a split-plot design, assigning a temperature to each incubator, then randomly assigning different growth mediums to the dishes inside individual incubators. Be careful here not to forget about replication. If you care about the effect of temperature at all, then you need replication with more than one incubator for each temperature. It is a very dangerous argument to claim that two incubators are absolutely identical, and so any differences in growth between them can only be due to their different temperature settings. Even if you bought them from the same supplier at the same time, it could be that the difference in their positions within the room means that one experiences more of a draught when open than the other, one of them may have antibacterial residues from a previous experiment that the other doesn't have, or one could have a faulty electrical supply that leads to greater fluctuations in temperature. Be very, very careful before claiming that two things in biology are identical (as they almost never are), and always replicate.

In a **split-plot design**, we have two factors and experimental subjects that are organized into a number of groups. For one factor (the **main-plot factor**) we randomly allocate entire groups to different treatment levels of that factor. We then randomly allocate individual experimental subjects within each of these groups to different levels of the other factor (the **sub-plot factor**).

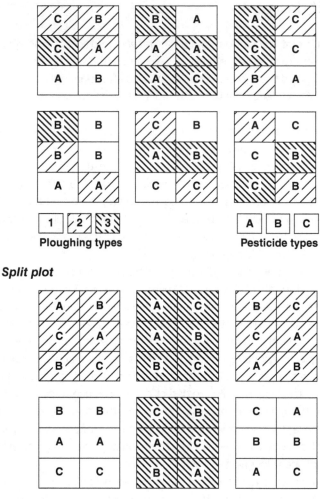

Figure 8.4 We have six fields each divided into six sections. We have two factors in our experiment: ploughing technique (with three levels: *1*, *2*, and *3*) and pesticide application (with three levels: *A*, *B*, and *C*). In the fully randomized treatment, we allocate four sections each to one of the nine (three times three) different combinations of the two factors. Allocation of a section to one of the nine treatment combinations is done entirely at random without consideration of which field a section is in. In the split-plot approach, entire fields are allocated a ploughing level. That is, two fields chosen at random are allocated to a given ploughing regime, and all the sections in a given field experience the same ploughing as each other. Ploughing is called the main-plot factor. We then take each field in turn and allocate the split-plot factor to individual sections within that field. We randomly allocate two sections within the field to each of the three methods of pesticide application. The split-plot approach may be attractive when it is practically very inconvenient to plough different parts of one field in different ways.

In both these cases the split-plot design is attractive because the spatial scales at which we can easily impose experimental variation differ between the two factors. A split-plot design can also be attractive if it is simply easier to impose variation in one factor than another. Imagine we want to explore the effects of three different levels of water temperature and four different lengths of previous fasting on the activity of crabs. We have at our disposal 48 crabs and a single large experimental aquarium. Each crab is randomized to a combination of water temperature and previous fasting period, and we have four replicate crabs for every combination. To help preserve independence of subjects we introduce crabs into the aquarium singly and observe them for thirty minutes each. If we completely randomize the order in which we assay crabs then we will frequently have to change the water temperature. This is inconvenient because it takes 24 hours to achieve this in our experimental system. To reduce this inconvenience we would split the crabs into six groups where all the crabs in a group have the same water temperature and there are two replicates of each fasting period per group. We then randomly order the eight crabs within each group, randomly order the six groups, and assay the 48 crabs in that order. Temperature has become the main-plot factor in a split-plot design: we have significantly reduced the number of times we will have to change the water temperature over the course of our experiment, and so shortened the duration of the experiment considerably.

➜ Split-plot designs should only really interest you in an experiment with multiple factors where one factor can be allocated easily to groups of subjects, but is practically awkward to assign to individual subjects.

Q 8.3 We argued earlier that replication is important. What would you do in the yeast-growth experiment described here if you only had access to three incubators and you wanted to explore three different levels of temperature?

8.5 Latin square designs

Like the split-plot designs just discussed, the Latin square design is another alternative to a fully randomized factorial design. The big attraction of the Latin square approach is in reducing the number of experimental units required. Suppose that we wanted to evaluate the quality of apples from orchards subject to four different pesticide regimes, so the factor of interest has four different levels. We require experts to evaluate the amount of insect damage in a box of apples taken from each orchard and estimate the percentage of apples that are fit for sale to humans. Since evaluation is time consuming, we require four experts. Further, to find sufficient space for our experiment, we used four different farms (each of which has several orchards). We might reasonably expect both the identity of the expert and the farm that the apples came from to impact on the score that a box gets, as well as any effect of pesticide treatment. Thus, since we have three factors in our study, each with four levels, we would ordinarily recommend that the number of subjects be a multiple of 64 (4³). This ensures that the experiment is balanced and has at least one subject for each combination of levels of the three different factors. Adoption of a Latin square design allows us to reduce this number of subjects considerably, such that we only need a multiple of 16 (4²). This is done by allocating pesticide treatments to orchards within farms and boxes of apples to expert evaluators according to an appropriate Latin square (see Figure 8.5). Such an approach can considerably reduce the size (and thus cost) of an experiment.

Some examples of Latin square designs

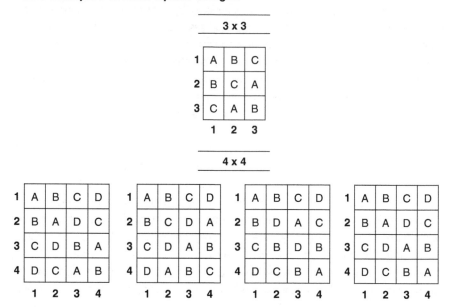

The first blocking factor (A) is represented by the columns and the second blocking factor (B) by the rows of these tables.

The treatment levels of factor C are represented by capital letters in the tables. Note the symmetry of these designs: each row is a complete block and each column is a complete block.

Figure 8.5 To make a Latin square, we first of all produce an $N \times N$ square grid, then place N distinct symbols (letters in our case), in the cells of this grid, such that each symbol appears once and only once in every row and every column (analogous to Sudoku puzzles). This grid is then used to allocate experimental units. For example, if we take the first (leftmost) of the 4 × 4 grids this could be used for the apple experiment described in the text. We want to investigate the effect of four pesticide treatments, that we label A, B, C, and D. We use four farms (represented by the rows 1 to 4 of the grid), and four experts (represented by the columns 1 to 4). Each expert will receive one box of apples from each of the different farms. Expert 1 will receive a box from farm 1 of apples grown under pesticide A, a box from farm 2 grown under pesticide B, a box from farm 3 grown under pesticide C, and a box from farm 4 grown under pesticide D. Expert 2 will receive a box from farm 1 of apples grown under pesticide B, a box from farm 2 grown under pesticide A, a box from farm 3 grown under pesticide D, and a box from farm 4 grown under pesticide C. In all, only 16 subjects (four orchards on each farm) will be required, whereas a traditional full-factorial experiment would have required 64 orchards.

However, there are restrictions on when Latin square designs can be applied. These designs are highly constrained in that the number of levels of each of the factors must be the same. For example, imagine we are interested in comparing the effects of three different diets on the growth rates of puppies. We are concerned that the breed of a puppy and the size of the litter it came from might be confounding factors. Now because diet has three levels (the three separate diets), then to use a Latin square design we need to impose that the other two factors also have three levels, so we would use three different breeds of dog and select subjects from three different litter sizes. This is quite a restriction. Let's say that to get the number of replicates you need, you have to use seven different breeds of dog. You could get around this problem if you could divide these seven breeds into three groups. We would only recommend this trick if there is a truly natural division (e.g. into gun dogs, lap dogs, and racing dogs, in our example). If you cannot achieve this symmetry, then you must use a fully randomized factorial design instead. An important assumption of the Latin square design is that there are no interactions between the three factors. In our example, we are assuming that the effects of breed, litter size, and diet act completely independently on the growth rates of puppies. Our knowledge of animal physiology suggests that this may well not be true. In cases like this, where you think interactions could be important, again we would recommend a factorial design instead, where interactions can be tested. However, in cases where you can justify the assumption of no interactions from your prior knowledge of the underlying biology, and where you can design a useful experiment so that all the factors have the same number of levels, the Latin square does present a very efficient design, in terms of reducing the numbers of replicates needed. In practice, we find that most scientists rarely find themselves in circumstances that justify use of this design.

→ The Latin square design is an alternative to a fully randomized factorial design using fewer experimental units. However, it can only be used to ask a very restricted set of scientific questions.

8.6 Thinking about the statistics

In any study, the experimental design that you choose and the statistics that you use to analyse your data will be tightly linked. Now that we have outlined some of the most common experimental designs that you will come across we want to finish this chapter by demonstrating this link with an example.

Suppose you are in charge of a study to evaluate whether a range of herbal extracts are of any use as antibacterial agents. To do this you decide to measure the growth of *Escherichia coli* bacteria on agar plates containing the different extracts, and also on control plates. You have nine extracts to test. How do you proceed?

Obeying our advice from Chapter 1, you decide to think about the statistical tests that you will use to analyse your data before you start collecting any. Your study has one factor: the different extracts. Thus, your study will be based around a one-factor design.

After discussion with a statistician (or reading a statistical textbook), you decide that the type of analysis appropriate for this study is a one-factor analysis of variance (or one-way analysis of variance). Do not worry if you have never heard of this test before, and have no idea how to actually do it. In fact, in these days of user-friendly computer programs, the actual doing of the test is the easy part—it is making sure your tests are appropriate for your data that can be tricky. The important thing from our point of view is that this test, as with all statistical tests, has a number of assumptions that must be met if it is to be used appropriately. One important requirement is that for each level of the factor you need at least two measurements (i.e. some replication). In your case this means that for every extract (and a suitable control), you need to make at least two experimental measurements of growth rate. A second requirement is that the data collected must be a measurement (i.e. measured on an interval or ratio scale, see Statistics Box 8.2 for details), not an arbitrary score. So if you chose to measure the average colony diameter of 100 randomly chosen colonies per plate, the analysis would be appropriate; but if instead you simply assessed growth on the plate on a 5-point scale based on categories (like 1 = no growth, 2 = very little growth, and so on), you could not use this analysis. There are several other requirements of this test, and whilst we will not go through them here one by one, that is exactly what you would do if you were in charge of this study. If you can meet all the requirements, all well and good, but what if you can't? First you need to consider the consequences of breaking a particular assumption. In general, failing to meet the requirements of a test will reduce your confidence in the result (and give ammunition to the Devil's Advocate). Failing to perfectly meet some assumptions may have little effect on the reliability of the test, and in this case you might decide to proceed with caution. However, this will often not be the case and failure to meet some requirements will entirely invalidate the test (which is not to say that you won't be able to get your computer to carry it out, just that the results would be untrustworthy). Detailed discussion of these different situations is well beyond the scope of this book, and if you find yourself in this situation we would strongly advise reading a good statistics book such as one of those in the Bibliography, or speaking to a statistician. If after this you decide that you cannot meet critical assumptions, you are left with two options: either to redesign the study to meet all the assumptions or to look for another suitable test with different assumptions that you can meet (bearing in mind that such a test may not exist, or may be considerably less powerful than your original choice).

Once you have decided on the test that you will use you can also begin to think about the power of the experiment you are proposing, and the kinds of sample sizes that will be required to make your study worth doing. Unless you know how you will analyse your data in advance it is impossible to make sensible estimates of power (since the power will depend on the test that you choose—see Chapter 6 for more on power).

Whilst these procedures might seem very long winded and slightly daunting, especially if you are not yet confident with statistics, we hope you can now begin to see clearly how thinking about the statistics when you think about your design, and long before you collect data, will ensure that you collect the data that allows you to answer your biological question most effectively.

STATISTICS BOX 8.2 **Types of measurement**

Different statistical tests require different types of data. It is common to split data into the following types:

- **Nominal scales**: a nominal scale is a collection of categories into which subjects can be assigned. Categories should be mutually exclusive but there is no order to the categories. Examples include species or sex.

- **Ordinal scales** are like nominal scales except now there is a rank order to the categories. For example, the quality of second-hand CDs might be categorized according to the ordinal scale: poor, fair, good, very good, or mint.

- **Interval scales** are like ordinal scales except that we can now meaningfully specify how far apart two units are on such a scale, and thus addition or subtraction of two units on such a scale is meaningful. Date is an example of an interval scale.

- **Ratio scales** are like interval scales except that an absolute zero on the scale is specified so that multiplication or division of two units on the scale becomes meaningful. Ratio scales can be either continuous (e.g. mass, length) or discrete (number of eggs laid, number of secondary infections produced).

Our advice is that you should try and take measurements as far down this list as you can, as this gives you more flexibility in your statistical analysis. Hence, if you are recording the mass of newborn infants then try to actually take a measurement in grams (a ratio measurement) rather than simply categorizing each baby as either 'light', 'normal', or 'heavy'; as this will increase the number and effectiveness of the statistical tools that you can bring to bear on your data. Of course, by taking our advice and deciding before data collection what statistical tests you will carry out, you will already know the type of data you require.

 Think about the statistics you will use before you collect your data.

■ Summary

- For factorial experiments, your analysis will be a lot easier if you have a complete design.
- Often hypotheses we are interested in testing involve the interaction between factors.
- If you are interested in such an interaction effect, then you must ensure that your experimental design and statistical analysis allow you to directly explore this interaction.
- Take care to avoid confusing levels and factors.

- Split-plot designs should only really interest you in an experiment with multiple factors where one factor can be allocated easily to groups of experimental units, but is practically awkward to assign to individual experimental units.

- The Latin square design is an alternative to a fully randomized factorial design using fewer experimental units. However, it can only be used to ask a very restricted set of scientific questions.

- Think about the statistics you will use before you collect your data.

Beyond complete randomization: blocking and covariates

9

Imagine we want to explore how a new drug affects blood pressure in humans, and we have a sample of people enrolled in our trial that we want to allocate to either the new drug or a placebo. There is nothing wrong with allocating individuals randomly to one of the two groups as discussed in Chapter 7. There are many confounding factors that might be linked to someone's blood pressure (e.g. age, sex, body mass index (BMI), diet, or occupation), but our randomization will average these out across the two groups. However, all these factors will add between-individual variation to our study, and so we may have to have quite a large sample size to get the statistical power that we want (see Chapter 6). If we don't have a good grasp of what confounding factors are likely to cause most of the between-individual variation (or the important confounding factors are difficult to measure on our subjects), then randomization or within-subject designs (discussed in Chapter 10) are our best approaches. However, if (for example), based on previous work, we expected that BMI might have a particularly strong influence on blood pressure, then alternatives might be open to us. We could reduce noise in our experiment by only using individuals from within a narrow range of BMI values, but this will affect how we can generalize from our study. If we only use subjects from the 'healthy' range of BMI values, then we lose our ability to inform about the effects of the drug on overweight and underweight people. Alternatively, we can use BMI as a blocking factor or covariate, and this chapter will introduce you to these techniques.

- We begin by explaining the concept of blocking in experimental design (Section 9.1).

- We then explore the range of things that you might be interested in blocking on (Section 9.2).

- We look at the potential costs of blocking to help you decide whether it is an attractive design for a given experiment (Section 9.3).

- We look at paired designs as a special form of blocking with only two subjects in a block (Section 9.4), and consider more generally how you might best select the size of blocks (Section 9.5).

- Finally, we introduce covariates (Section 9.6), and consider interactions involving covariates (Section 9.7).

9.1 The concept of blocking on a particular variable

If a given variable is expected to introduce significant variation to our results, then we can control that variation by **blocking** on that variable. This involves splitting our experimental subjects up into blocks, such that each block is a collection of individuals that have similar values of the blocking variable. We then take each block in turn and for each block distribute individuals randomly between treatment groups. It is thus a more complicated alternative to complete randomization. Blocked designs are sometimes called *matched-subject designs*, and hence blocking can be called *matching*.

 See Section 9.1 of the supplementary information for more on allocation strategies, including explanation of the terms *restricted random allocation*, *block allocation*, *permuted block allocation*, and *block randomization*. Go to: **www. oxfordtextbooks.co.uk/orc/ ruxton4e/.**

Suppose that we are interested in whether the type of food we give to a greyhound affects its running speed. We have 80 greyhounds and want to test the effects of four food types. The 80 greyhounds are of different ages, and age is expected to have a strong effect on running speed. We could treat age as a **blocking** factor in the experiment.

First of all, we rank the dogs by age, and then partition this ranking so as to divide the dogs into blocks, so that those in a block have a similar age. This might be 20 blocks of 4 dogs, 10 blocks of 8, or 5 blocks of 16. This choice does not matter too much (but see Section 9.5 for further thoughts on selecting block size), providing the number of individuals in a block is a multiple of the number of treatment groups. We then take each block in turn and randomly allocate all the dogs from one block between the treatment groups, in exactly the way that we did in the fully randomized design. We repeat the same procedure with all the blocks. This approach of using blocks (sometimes called strata) as part of our allocation procedure is called *stratified block allocation* and contrasts with the complete randomization that we described in Chapter 4.

So what does this achieve? One way to think of it is this: in the randomized experiment we will find ourselves comparing an old dog in one treatment and a young dog in another. If we compare their running speeds and find a difference, part of this may be due to food, and part due to age (and the rest due to random noise). In contrast, in the blocked design we are making comparisons within our experimental blocks, so we will only ever be comparing dogs of similar ages. With differences due to age much reduced in any given comparison, any effects due to diet will be much clearer (see Figure 9.1). The blocking has effectively controlled for a substantial part of the effects of age on running speed. You can think of this blocked design as a two-factor design, with food type as the first factor, and age block as the second.

Effectively what we are doing is dividing the between-individual variation into variation between blocks and within blocks. If we have blocked on a variable that does have an important effect on between-subject variation, then the between-block differences will be much greater than those within blocks. This is important, because in our final statistical analysis we can use 'blocks' as an additional factor in our ANOVA. This will allow us to test for between-treatment differences within each block (where between-subject differences are small) rather than across the whole sample (where between-subject differences would be larger). Or, as a statistician might say, we can control for variation amongst blocks before looking for differences due to treatment. Thinking again in terms of the questions our design allows us to answer, our randomized block design allows us to answer the following question:

Do food types affect running speed in greyhounds, once age has been controlled for?

Or put in a slightly different way:

Are age-corrected running speeds in greyhounds affected by the food types?

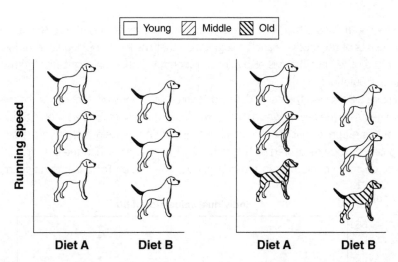

Figure 9.1 The left-hand panel illustrates a fully randomized design to explore the effect of diet on the running speed of dogs. Dogs are simply randomly allocated to one of two groups: one group experiences diet *A*, the other diet *B*. However, if we know the age of the dogs and expect age to have a strong effect on running speed, then we might instead opt for the blocked design illustrated in the right-hand panel. Here dogs are first allocated to three blocks on the basis of their age (young/middle/old). Each of these age blocks is taken in turn and the dogs within that block randomly allocated to either diet *A* or diet *B*. Subsequent analysis will then test for a difference in running speed due to diet between the two groups of dogs of each given age block. If a large proportion of the within-group variation in the fully randomized experiment was due to variation in age, then the blocked design may be more effective at testing for an effect of diet.

And (as should be no surprise to you by now) if we can remove some of the inherent variation from our study, we can increase our statistical power. You can find more on this in Box 9.1.

BOX 9.1 Variation, blocking, and power

Imagine that you are interested in whether the 'song' that a male fruitfly produces with his wing during courtship differs in its characteristics between populations of a given species. You collect individual males from two populations, and measure a particular component of their courtship song, the inter-pulse interval (IPI). This character is easy to measure and known to vary between species. You plan to measure 15 males from each population, but the procedure that you use to measure IPI is time consuming, and realistically you can only measure ten flies a day, and so you are going to have to measure flies over three days. Previous work has shown that courtship song can be affected by factors like atmospheric pressure which are likely to vary between days. Sensibly you decide on a randomized block design where you measure five flies of each population on each day. Figure 9.2a shows the results of the study. There is a suggestion that the IPI of one population is lower than the other, but there is also a lot of random variation amongst flies from the same population.

Figure 9.2b shows the same data plotted in a different way, separating the measurements by the day on which they were made. The first thing that should be clear from this graph is that there is much less variation between flies from the same population when measured on the same day than there is when the data from the three days are pooled (as in the first panel). The reason for this can be seen by

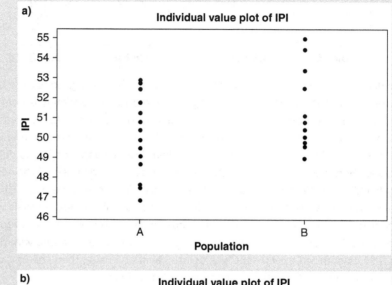

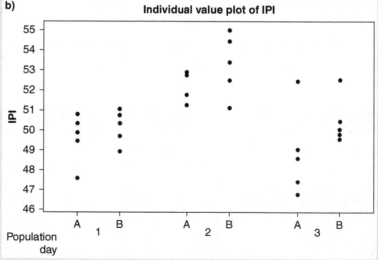

Figure 9.2 a) The IPI of individual flies from the two populations and b) the same data now broken down by day of measurement.

looking at the average IPI of the flies on each day. Overall flies on day 2 seem to sing slightly faster (i.e. with a smaller IPI) than flies on day 1, and the flies on day 3 may sing slightly slower. This day-to-day variation is part of the random variation in the pooled data, and is part of the reason that the data is so noisy. If we use *Day* as a blocking factor in our statistical analysis, then we are effectively removing the day-to-day variation, making any differences between populations easier to see. Or in statistical jargon, we are looking for population differences having controlled for day-to-day variation.

Imagine that we had carried out a power analysis prior to carrying out this study to estimate the power of our planned study to detect a difference in IPI of 1.5. The power of the study using the variation in 9.1a as our estimate of the random variation (that is, carrying out the statistical analysis without the blocking factor) would have given a power of 0.55, whilst the analysis including the blocking factor and using the within-day variation would give a power of 0.84. So if we ignore the effect of the blocking factor, we would expect to detect a difference of 1.5 only just over half of the time, but if we take the blocking factor into account, we increase our chance of detecting the same difference to over 80% of the time.

➡ If you know and can measure some aspect of experimental subjects that is likely to explain a substantial fraction of between-subject variation then it can be effective to block on that factor.

9.2 Blocking on individual characters, space, and time

You can block on any factor (indeed on any number of variables) that you think might be important in contributing to variation between individuals that have been given the same experimental treatment. The only condition for a characteristic to be used as a blocking factor is that you can measure it in individuals, so you can allocate them to groups of individuals that are similar in that trait (blocks). We will refer to this as *blocking by individual characters*.

There is a second type of blocking that is commonly used in biological experiments that we will call *blocking in space*. Imagine that in our tomato experiment discussed in Chapter 8 we find that we can't fit all our plants into a single greenhouse, but instead need to use three different green houses. If there are differences between the conditions in the greenhouses that affect growth rates, then by carrying out the experiment at three different sites we are adding some noise due to site differences. We could simply ignore this variation and use a fully randomized design where every plant is allocated at random to a greenhouse irrespective of its treatment. However, this could be a very inefficient way to carry out such an experiment. We cannot simply have all of the individuals in a given greenhouse getting the same treatment, as this would introduce greenhouse identity as a confounding factor. A common solution to this problem is to block by greenhouse. What this means is that we allocate equal numbers of each treatment to each greenhouse, and note down which greenhouse a plant is grown in. We now have a two-factor design with feeding rate as the first factor, and greenhouse as a second 'blocking' factor. This means we can make comparisons between plants in the same greenhouse, comparisons that will be unaffected by any differences between greenhouses. Blocking by site is frequently seen in large agricultural trials (where several fields or even farms may act as the blocks), but could equally apply to any experiment where assays are carried out at different sites. Sites can be anything from fish tanks, to shelves in a growth cabinet, to labs in different countries.

The final kind of blocking that we will discuss is *blocking in time*. This is exactly analogous to blocking in space except that the measurements are carried out at different times rather than in different places. In our tomato example, if you only had one greenhouse and had to carry out the experiment in three batches to get the sample sizes you needed, then you could block on batch number (i.e. blocking in time) in an exactly analogous way to the situation we described of treating each of three greenhouses as a block (blocking in space). This may also be useful if you can't possibly assay your whole experiment in a single time period, and are concerned that the time when a particular experimental subject is measured might influence the measurement taken from it (see Chapter 11 for more on this).

The recipes described previously, where equal numbers of subjects from each block are assigned to all of the treatment types (or equal numbers of each treatment type are assigned to each block), are called *balanced complete-block* or *randomized complete-block* designs. It is possible to have different numbers of subjects within a block assigned to different treatment groups, or indeed to have treatment groups that do not get experimental subjects from every block. Such incomplete-block designs are much harder to analyse statistically. If you are forced into such a situation, perhaps by a shortage of experimental subjects, then our advice is to consult a statistician beforehand.

 As well as intrinsic characteristics of subjects, it can also sometimes be useful to block on time and/or space.

 Q 9.1 Imagine that two scientists need to evaluate the quality of tomatoes grown in an experiment with three different greenhouses and four different treatment types. They are worried that the score a tomato gets might be affected by which of them measures it; can blocking come to their aid?

9.3 The advantages and disadvantages of blocking

The advantage of blocking is in reducing the effect of intrinsic between-subject variation so that treatment effects are easier to detect. As such, it can potentially increase our statistical power. There are drawbacks too. Blocking using a characteristic that does not have a strong effect on the response variables that you eventually measure in order to compare treatments is worse than useless. Hence, do not block on a characteristic unless you have a good biological reason to think that the characteristic that you are blocking on will have an influence on the variables that you are measuring. The reason that blocking with an ineffective variable actually reduces your chances of finding a treatment effect is because the statistical test loses a bit of power when you block, because that power is used in exploring between-block differences. Normally this price is worth paying if blocking leads to reduced between-individual variation within blocks; but if it doesn't, you lose out.

Blocking also cannot be recommended if you expect high drop-out rates such that some treatment groups end up with no representatives from several blocks. This effectively produces an unplanned incomplete-block design, which, as we discussed earlier, can be very hard to analyse statistically. Generally, the benefits of blocking increase as sample sizes increase, hence we would not recommend blocking in situations where you are constrained to using rather fewer experimental subjects than you would ideally like. In such cases, complete randomization is usually better.

 Don't block on a factor unless you have a clear expectation that that factor substantially increases between-individual variation.

Q 9.2 You have a sample of student volunteers on which you intend to explore the effect of four different exercise regimes on fitness. Are there any variables you would block on?

If you are concerned that blocking seems to violate the concept of independence of samples discussed in Chapter 5 see Section 9.3 of the supplementary information. Go to: **www.oxfordtextbooks.co. uk/orc/ruxton4e/**.

9.4 Paired designs

A common form of blocking is a **paired design**. For example, imagine that we want to look at the effect of an antibiotic injection soon after birth on the subsequent health of domestic cattle calves. We might use sets of twins as our experimental pairs, randomly assigning one calf from each set of twins to get the antibiotic. The attraction of this is that we can perform statistical tests (e.g. paired t-tests or z-tests) that compare within pairs. This has the same benefits as all forms of blocking: eliminating many potential confounding factors, because the two twin calves in a pair will be genetically similar, will be fed by the same mother, and can easily be kept together in order to experience the same environmental conditions. But notice one drawback: since we only perform our investigation on calves that were from a set of twins, can we be sure that our results will also apply to calves that were born alone? This type of extrapolation concern was dealt with in Section 6.4.1. Paired designs need not only use sets of twins. For example, a study looking at sex differences in parental feeding in seabirds might compare the male and female partners within a number of breeding pairs. Comparing within pairs controls for variation due to factors such as number of chicks in the nest and local foraging environment, both of which might be expected to contribute considerably to variation in feeding behaviour.

In **paired designs**, we divide the population into pairs of similar individuals (or use naturally occurring pairs) and randomly assign the individuals of each pair one to each of two treatment groups.

Paired designs can be attractive for similar reasons to blocking generally, but be careful to think about whether your choice of pairs restricts the generality of conclusions that you draw from your sample.

9.5 How to select blocks

In Section 9.1, we said that it did not matter much how you divided subjects into blocks. For example, if you had 80 subjects and four treatment groups, you could go for 20 blocks of 4, 10 blocks of 8, or 5 blocks of 16. Let's see if we can give you some clearer advice.

First of all, don't have any more blocks that you can justify if you are blocking on something that must be scored subjectively. For example, say the subjects are salmon parr and you are blocking them according to an expert's opinion on their state of health from external examination: given as a score on a 1–20 scale. In this case, it might be hard to convince yourself that those in the 18th and 19th blocks out of 20 are really different from each other. That is, it may be that the subjectivity involved in scoring fish on this fine scale adds noise to your study. If this is a valid concern, then you should use fewer blocks.

If you are ranking on something more objective, like the length of the parr, then you may as well go for as many blocks as you can, in order to reduce variation within a block due to length. However, our interpretation of the analysis of blocked experiments can be complicated if there is an interaction between the experimental factor and the blocking variable. For example, if we are interested in how quickly salmon grow on four different diets, straightforward interpretation assumes that although diet and initial length both affect growth rate, they do so separately. To put this another way, it assumes that the best diet for large fish is also the best diet for small ones. This may not be true. In order to explore whether it is or not, you need several fish on a given diet within each block. Hence, if you are concerned about interactions, and in general you often will be (see Chapter 8), then adopting 10 blocks of 8 would be a better choice than 20 blocks of 4.

 Make sure you have replication of a treatment within a block if you are interested in interactions involving blocking factors.

9.6 Covariates

Let's return to our experiment where we compare the running speed of greyhounds fed different diets. Recall that we expect that differences in age will be an important source of variation between subjects' running performances, and the approach we took previously was to block by age categories. An alternative way to deal with between-subject variation due to age is to record the age of each dog alongside its speed. In our statistical analysis, we could then attempt to control for the effect of age, and so make any difference between the diets more obvious. In other words, by statistically removing some of the between-individual variation, we have increased the power of our experiment to detect any treatment effects. Box 9.2 shows how this works. This design allows us to answer the same question as the blocked design, namely:

Do food types affect running speed in greyhounds once age has been controlled for?

Whether blocking by age or treating it as a covariate is the best way to proceed will depend on how the age is expected to affect running speed (see later in this section).

9.6 Covariates **135**

BOX 9.2 How controlling for a covariate can increase the power of the study

Suppose you are interested in whether treating a crop pest species of beetle with a pathogenic fungal agent has the potential to reduce their reproductive output. You carry out a simple fully randomized one-factor experiment. You allocate ten individual females to each treatment (fungus infection and control) and measure how many eggs they lay before they die. Figure 9.3a shows the kind of data you might get from such a study. Whilst there is a suggestion that there may be an effect of the treatment, it is clear from the figure that there is an awful lot of variation within each of the treatment groups, and this noise is going to make any effects hard to detect. Suppose that we carried out a power analysis prior to the study, using the amount of noise in this data as our estimate of variation, the power of the proposed study to detect a difference of, for example, ten eggs would be low (about 30%.)

However, we know that for many insect species the lifetime fecundity is strongly influenced by their body size, and so variation in body size within treatment groups may be accounting for some of this variation in fecundity. Figure 9.3b shows that our intuition was correct in this case; there is an obvious relationship between a female's weight and the number of eggs she lays in our study. And this allows us to control for this variation. Essentially, how this works is as follows: we fit a line through our data that relates the number of eggs laid to the weight values. One way to think of this line is as providing the predicted fecundity for females of any size. We can then use this line to ask, for any individual in our study, how many more or fewer eggs they produced than would be expected for a female of their size.

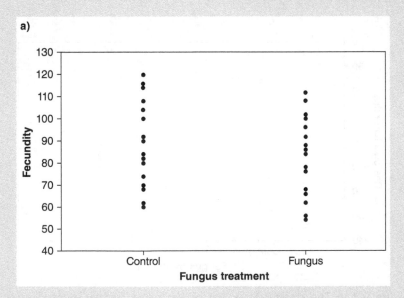

Figure 9.3 a) The lifetime fecundity of the females in the control and treatment groups.

This amounts to asking how far above or below the line does their actual fecundity value fall? Statisticians call this their residual fecundity; we can think of it as a measure of their fecundity corrected for their weight. Figure 9.3c shows these weight-corrected values plotted in the same way as the actual values in 9.3a. What is clear is that there is much less variation within treatments now that we have controlled for weight. For comparison, had we carried out a power analysis using the variation in this figure as our estimate of the amount of random variation, the predicted power of our study would have risen to close to 100%.

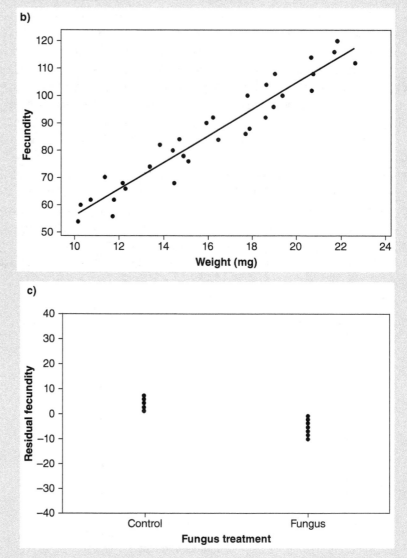

Figure 9.3 b) The lifetime fecundity of the same females plotted against their weight, and showing the line of best fit through the data. c) The residual fecundity of females after controlling for their weight.

As always we need to think about the statistics that we will ultimately use. In this case (where we measure a covariate), the design leads naturally to an ANCOVA (analysis of covariance) rather than an ANOVA. There are three requirements for such an analysis of covariance that are important to bear in mind.

The first requirement is that the covariant is a *continuous variable* like age. Continuous variables (such as height, weight, or temperature) are measured, can take a great range of values, and have a natural order to them (for example, 3.2 seconds is shorter than 3.3 seconds, and 5 metres is longer than 3 metres). In contrast, discontinuous (or discrete) variables involve a set of categories. They may only take limited values, and need not (but can) have any natural order to them. Typical discontinuous variables might be the colour of a seaweed, classified as *red*, *green*, or *brown* (no natural order), or the size of a tumour, classified as *small, medium*, or *large* (natural order). The distinction is not always clear cut, whilst classifying dogs as *young, middle-aged*, or *old* treats age as a discontinuous variable; the actual age of a dog in years might be regarded as discrete if only measured to the nearest year, but continuous if we allow fractions of years. If you need a reminder about different types of variable, have a look at Statistics Box 8.2. In our study, if we knew the exact age of our dogs then we could treat age as a covariate. But if we could only classify them as either *young, middle-aged*, or *old* then we should treat age as a blocking factor.

The second important assumption of ANCOVA is that we expect some form of linear relationship between running speed and age, so that speed either increases or decreases linearly with age. If we actually expected speed to increase between young and middle-aged dogs, but then decrease again as dogs become old, or to follow some other non-linear relationship, then a blocked design will be far more straightforward to analyse. Whether age is treated as a blocking factor or a covariate, your statistical test will be most powerful if the age distributions of dogs on the two diets are similar.

The third important assumption of ANCOVA is that any relationship between our continuous covariate and running speed is the same for our different treatment groups. In statistical terms, it requires that there is no interaction between our factor and our covariate. We will discuss this at greater length in the next section.

There is no reason to restrict yourself to one covariate. As a general rule, if there are characteristics of samples that you can measure and that you think might have a bearing on the response variable that you are actually interested in, then measure them. It's much better to record lots of variables that turn out to be of little use, than to neglect to measure a variable that your subsequent thinking or reading leads you to believe might have been relevant (but bear in mind our advice on focused questions in Chapter 2). Once your experiment is finished, you can't turn the clock back and collect the information that you neglected to take first time around.

➜ Recording covariates can be a useful way to account for noise from measurable confounding factors, but blocking can often be more powerful unless your system satisfies two important assumptions about linearity of response and the continuous nature of the covariate.

9.7 Interactions between covariates and factors

In Section 8.2 we discussed interactions where the effect of one factor depends on the level of another factor. However, another important type of interaction that you will come across is between factors and continuous variables (covariates). An interaction between a factor and a covariate means that the effect of the factor depends on the level of the covariate. Figure 9.4 shows situations with and without such interactions. Both panels show the results of experiments to look at the effect of the brightness of a male stickleback's mating coloration (measured under standardized conditions) on his mating success when females are exposed to the males. Our experiment randomizes males to one of two experimental treatments: a tank illuminated by normal broad-spectrum light or a tank illuminated by a narrower band of frequencies giving red light. The two things that we expect to affect mating success are lighting treatment (a factor with two levels: normal or red) and a covariate (investment in pigmentation by the male, measured as a continuous variable: brightness). In Figure 9.4a, there is no interaction. Mating success is higher in normal light, but the lines are parallel, and this indicates that the effect of light type is the same at all levels of male brightness. Contrast this with Figure 9.4b. Here we see the relationship between mating success and brightness differs under the two lighting regimes (i.e. there is an interaction). The effect of light type now depends on the brightness of the fish concerned, with normal light increasing mating success of bright fish, but having no effect (or perhaps a slightly negative effect) on dull fish. Hence, you should be aware that it is possible to design experiments that allow you to test for interactions between a covariate and a factor as well as between two factors. In general, this will require that the values of the covariate overlap to some extent for the different levels of the factor. So we would want to avoid an experiment where all the fish tested in normal light had high levels of brightness, but all those tested in red light have low levels of brightness. Similarly, as we have already mentioned, if we want to use covariates to control for differences between individuals, such techniques are only really very useful (and definitely only straightforward) if there is a linear relationship between the thing that we are measuring and the covariate and if this relationship is the same for our treatment groups (there is no interaction between the covariate and the treatment).

 Covariates and discrete factors can interact; and such interactions will often require plotting the data to interpret effectively.

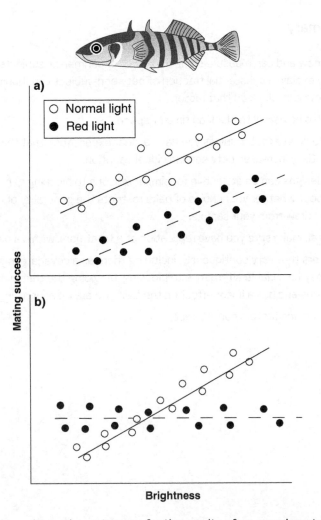

Figure 9.4 Two alternative outcomes for the results of an experiment to explore the effects of both a factor with two levels (normal or red lighting) and a covariate (brightness of the fish's exterior) on a male fish's mating success. In (a) both variables have effects but there is no interaction between these effects. That is, we can see that males are more successful if they are intrinsically brighter and/or if they are seen by females under full-spectrum normal light. But importantly there is no interaction: the effect of lighting regime is the same for all fish, no matter their brightness. In (b) there is an interaction. That is, we cannot say anything general about the effect of light without referring to a specific brightness level and we cannot say anything about the effect of brightness without referring to a specific lighting regime. Specifically, it seems that higher brightness is advantageous under normal lighting, but that brightness is immaterial to mating success under red light only. Another way to describe the interaction shown is that very bright individuals are more successful under normal lighting, whereas individuals of low brightness have less success under normal lighting than they do under red light.

■ **Summary**

- If you know and can measure some aspect of experimental subjects that is likely to explain a substantial fraction of between-subject variation then it can be effective to block on that factor.

- It can also be useful to block on time or space.

- Don't block on a factor unless you have a clear expectation that that factor substantially increases between-individual variation.

- Paired designs can be attractive for similar reasons to blocking generally, but think about whether your choice of pairs restricts the generality of conclusions that you draw from your sample.

- In general, make sure you have replication of a treatment within a block.

- For factors that vary continuously, including them as a covariate in your study may be a viable alternative to blocking, providing the trait in question is continuous and has a linear effect on the trait you are measuring in your study.

- Covariates and factors can interact.

Within-subject designs

10

In this chapter we consider designs where, rather than comparing between subjects, we compare measurements on the same subject when that subject is exposed to different treatments.

- We begin by carefully defining how within-subject designs work (Section 10.1).
- We then consider both the advantages (Section 10.2) and disadvantages (Section 10.3) of this approach relative to the between-subject designs we have considered until now.
- We explain why within-subject designs need not fall foul of pseudoreplication (Section 10.4), but do warn that they can take a long time to complete (Section 10.5).
- We consider in which order you might want to apply the treatments to different subjects (Section 10.6).
- We also consider the apparent similarity between within-subject designs and randomized block designs (Section 10.7).
- We finish by considering how within- and between-subject elements can be combined within a single design (Section 10.8).

10.1 What do we mean by a within-subject design?

The experimental designs we have discussed so far in this book have involved comparing between subjects in different groups, each group being assigned a different treatment. It is no surprise that these designs are sometimes called *between-subjects* or *between-groups* or *independent measures* designs. The alternative is a **within-subject design** in which an individual subject experiences different treatments sequentially and we compare between measures of the same individual under the different treatments. We have already come across one within-subject design in the context of longitudinal studies (Section 5.6). Here we explore them in an experimental context.

> In a **within-subject design** (sometimes called a *cross-over* or *repeated-measures design*), experimental subjects experience the different experimental treatments sequentially, and comparisons are made on the same individual at different times, rather than between different individuals at the same time.

In a within-subject design experimental subjects experience the different experimental treatments sequentially, and comparisons are made within the same individual.

10.2 The advantages of a within-subject design

Imagine that you want to investigate whether the material offered to hens as lining for their nestboxes influences how many eggs are laid. You have a number of henhouses, each of which has a number of hens that are free to walk around outside the henhouse but prevented by fencing from moving between henhouses. A number of different nest lining materials could be explored: such as hay, wood shavings, sawdust, shredded paper, or commercially produced rubber matting. You could do a simple randomization experiment, taking a number of henhouses and randomly allocating houses to one of the different lining materials. However, simple randomization is going to require you to have a lot of henhouses. This will be particularly true if henhouses tend to vary in positioning or construction so that we expect inter-house variability to be high.

Q 10.1 Would you be interested in blocking in the henhouse experiment? If so, what factor would you block on?

One potential solution to this problem is to use henhouses as their own controls, so that you are comparing the same house under different treatments. Let's consider the simple case where the farmer has been using wood shavings but is shortly to lose their supplier for this and wants to look at alternatives; you are comparing rubber matting and hay. You should randomly allocate houses to one of two treatment sequences. The first receives rubber for three weeks then hay for three weeks. The other receives hay for three weeks then rubber for three weeks. You can then statistically compare the numbers of eggs laid in the same house under the two different regimes, using a paired *t*-test. Because you are comparing within (rather than between) houses, between-house variation is much less of a concern, and you should be able to get clearer results than from a fully randomized design with the same number of houses. Figure 10.1 compares some of the different designs that could be used to address this kind of question.

 Comparing within subjects removes many of the problems that between-group designs experience with noise.

10.3 The disadvantages of a within-subject design

There are two potential drawbacks to within-subject designs: period effects and carry-over effects. The first of these can be handled with careful experimental design; the second is more of a challenge.

10.3.1 Period effects

Notice that in the henhouse experiment we need two groups: one that experiences rubber first then the hay, and another that experiences the two treatments in the opposite order. We need two groups to avoid time being a confounding factor. If the laying

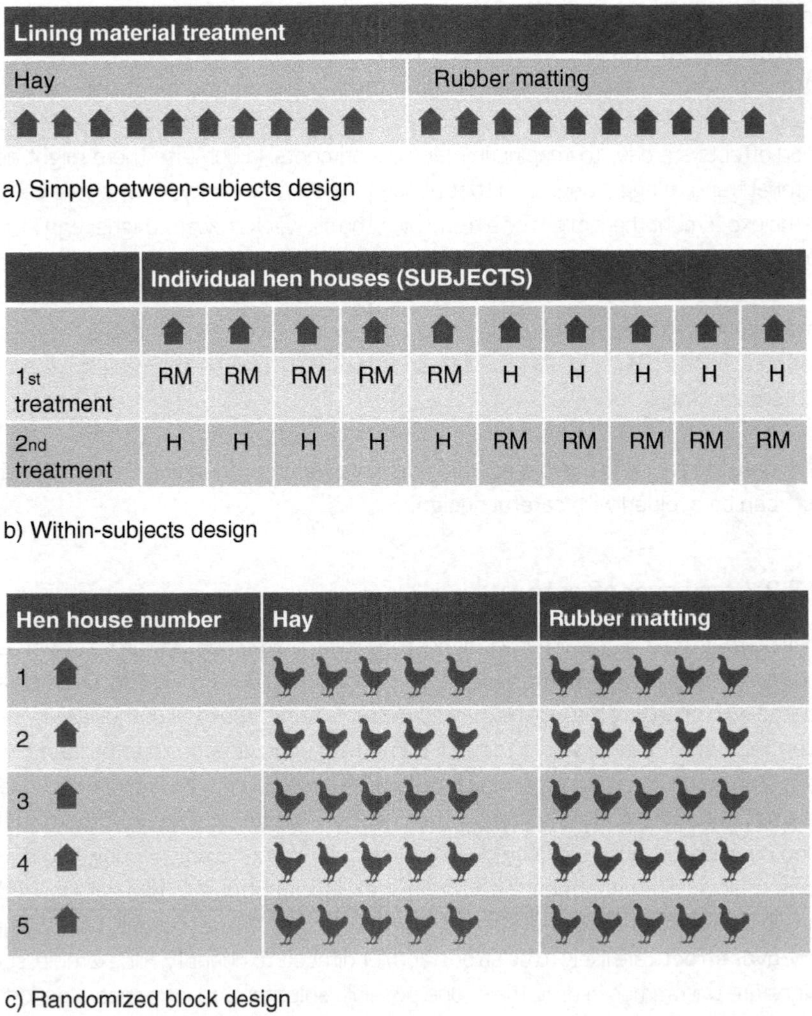

Lining material treatment	
Hay	Rubber matting
🏠🏠🏠🏠🏠🏠🏠🏠🏠🏠	🏠🏠🏠🏠🏠🏠🏠🏠🏠🏠

a) Simple between-subjects design

	Individual hen houses (SUBJECTS)									
	🏠	🏠	🏠	🏠	🏠	🏠	🏠	🏠	🏠	🏠
1st treatment	RM	RM	RM	RM	RM	H	H	H	H	H
2nd treatment	H	H	H	H	H	RM	RM	RM	RM	RM

b) Within-subjects design

Hen house number	Hay	Rubber matting
1 🏠	🐔🐔🐔🐔🐔	🐔🐔🐔🐔🐔
2 🏠	🐔🐔🐔🐔🐔	🐔🐔🐔🐔🐔
3 🏠	🐔🐔🐔🐔🐔	🐔🐔🐔🐔🐔
4 🏠	🐔🐔🐔🐔🐔	🐔🐔🐔🐔🐔
5 🏠	🐔🐔🐔🐔🐔	🐔🐔🐔🐔🐔

c) Randomized block design

Figure 10.1 Three ways to design a henhouse study. In the simple between-subjects design (a), individual henhouses are the subjects and are randomly allocated to one of the two lining material treatments. In the within-subjects design (b), the henhouses are again the subjects but they receive both treatments, one after the other, with half receiving rubber matting (RM) first and the other half receiving hay (H) first. In the randomized block design (c), individual hens lay in separate enclosures within each henhouse. Each henhouse receives both treatments simultaneously, with treatments being randomly allocated to individual enclosures within them; so in this design the individual hens are the subjects, while the henhouses become blocks.

rate of the hens changes over time for some reason unrelated to nesting material, then this would confound our results if all the houses were exposed to the rubber first followed by the hay treatment. By making sure that we have an equal number of henhouses experiencing all possible orderings of the treatments (there are only two orderings in our case) we are **counterbalancing**. Counterbalancing is an effective way to avoid time

Counterbalancing in a within-subject design involves allocating at least one subject to every possible sequence of the treatments. If there are N treatments then the number of sequences is $N! = N*(N-1)*(N-2)* \dots *1$.

Q 10.2 What if (rather than having the two counterbalanced groups suggested earlier) we had one group that experienced the experimental treatment in one-half of the time and the control treatment in the other, and another group that experienced the control treatment throughout. Wouldn't this allow us to check for period effects?

becoming a confounding factor, but may be impractical if you have a large number of treatments (more than four); see Section 10.6.

A *period effect* then is a systematic difference between treatment periods in the measured variable of interest regardless of any effect of treatments applied. Such period effects are easy to imagine in many experiments. In our case there might be a seasonal trend to egg laying, such that at the particular time of year when we carried out the experiment there might be a trend over the six weeks towards higher egg laying. To combat period effects it is good to have a design that is 'uniform across periods'. All this means is that when we look across sequences each treatment appears the same number of times in each period. In our hay-versus-rubber case this is true, each treatment appears once in each period, and this will always be the case if we adopt counterbalancing.

Care is required to avoid introducing time as a confounding factor, but this can be avoided with careful design.

10.3.2 Carry-over effects

It may be that the effect of a particular nestbox lining on the hens' laying rates persists for days after that lining is removed. Alternatively or additionally, any disturbance associated with changing the regime may adversely affect the birds. For these reasons it is worth looking at your raw data to see if something unusual appears to be happening at the start of each regime, after which the subjects settle down to a more consistent pattern of behaviour. If this is occurring, then it may be prudent to discard data from the period immediately after a changeover. For example, in your chicken study, you might decide to apply each treatment as a three-week block but only to use data from the last two weeks of each block. Within-subject designs are not particularly useful when **carry-over effects** are likely to be strong and/or difficult to quantify. For example, such designs are uncommon in drug trials. One possible solution is to introduce a period of normal conditions between the two regimes, to minimize carry-over effects; these are often called *washout periods*. In our hay-versus-rubber nestbox lining trial this might involve returning the chickens to the lining used before the start of our study (wood shavings) for a week between treatment regimes. Although this costs you an extra week of husbandry, it may avoid you having to delete data points from the first week of a treatment period as unrepresentative.

Carry-over effects occur when a treatment continues to affect the subsequent state of experimental subjects, even after that treatment is no longer applied.

If carry-over effects are trivial then we have the ideal situation of *reversibility*, where at the end of every treatment-and-measurement phase the subjects revert back to the state they were in at the start of that phase. Thus we can reasonably consider that experimental subjects' response to one treatment will be unaffected by any history they have of previous treatments. The within-subject design for the henhouse experiment only works because we can measure non-destructively (we don't have to harm a bird to measure how many eggs it has laid). If the question was whether nestbox lining increases the wall thickness of the heart, then chickens would probably have to be killed in order to be measured, and such a cross-over design would not be practical.

Similarly, it would not be practical to compare the effectiveness of different teaching techniques on how quickly young children develop reading skills using a within-subject design. Once the child has learned to read using one technique we cannot wipe away their reading ability so as to measure how well they re-learn under the second technique. Once they have learnt the first time, we cannot take that away. So comparison of the teaching techniques would require a between-subject design rather than a within-subject design.

There is a lot of jargon associated with carry-over effects. Some texts will differentiate between two different types of carry-over effects: *sequence effects* and *order effects*. Imagine we have three treatments: A, B, and C. If a subject's response to treatment C is affected by whether or not it has already experienced treatment A, regardless of how long ago it experienced treatment A, then this is an *order effect*. If, however, it matters whether A came directly before or whether there was an intervening treatment (treatment B), then this is a *sequence effect*. Similarly if carry-over effects only influence the treatment in the next period, but not in any subsequent periods this is called a *1st-order carry-over effect*, longer-lasting carry-over effects are called *higher-order carry-over effects*. Some texts will recommend *balanced designs* (where every treatment precedes every other treatment the same number of times—normally once), and *strongly balanced designs* (where every treatment precedes every treatment including itself the same number of times—usually once) as good designs for allowing carry-over effects to be detected and controlled for statistically. A counter-balanced design is balanced; and if we add another time period so that the final treatments of our counterbalanced design runs over two time periods then we get a strongly balanced design. Figure 10.2 shows examples of these designs. Other designs that have desirable properties in detecting and controlling for certain types of carry-over effects, but do not require as many different sequences as counterbalancing, are available, but we are not going to explore these in any detail since the associated statistical analysis is far from straightforward. Our advice is that where you (or the Devil's Advocate) have reason to be concerned about carry-over effects that you cannot negate using washout periods then you are better to switch to a between-subjects design. If you really want to carry out a within-subject experiment where you cannot dismiss concerns about carry-over

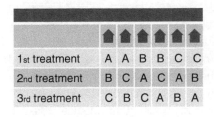

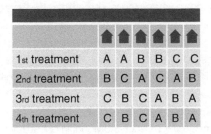

Counter-balanced design   Strongly balanced

Figure 10.2 Counter-balanced and strongly balanced within-subject designs for three treatments, A, B, and C.

effects then you should probably consult a statistician before embarking on it, but we provide some references that will help you get the most from that consultation.

→ Within-subject designs are not suitable for situations where a subject cannot be returned to the condition it was in before the treatment started. Make sure that you leave sufficient time for a treatment to stop having an effect on subjects before applying the next treatment and taking measurements of the effect of the new treatment.

10.4 Isn't repeatedly measuring the same individual pseudoreplication?

Within-subject designs necessarily involve experimental subjects being measured more than once—why does this not suffer from the problems of pseudoreplication discussed in Chapter 5? The important difference is that in a within-subject design a subject is measured more than once, but there is only one measurement made of each individual under each experimental condition (in the henhouse case, once with rubber and once with hay). We then compare between the measurements within the same individual. The problem of pseudoreplication only arises if we measure a subject more than once under the same conditions, and then try to use these data to compare between individuals under different conditions. If you are finding this distinction confusing, Box 10.1 explores the idea further.

BOX 10.1 Within-subject designs and pseudoreplication

If you are still having problems with the idea of pseudoreplication and within-subject designs, you might find the following helps your intuition. Imagine we want to compare the running speed of greyhounds on different diets again. We have 20 dogs and we measure the speed of each twice, once after a month on the standard diet, and once after a month on a high-protein diet (obviously some dogs get the standard diet first, whilst others get the high-protein diet first). This means we have twice as many measurements as we have dogs—surely this is pseudoreplication? To understand why it is not, you need to think about the real object of this study.

   Whilst we have two measures of speed for each dog, what we are actually interested in is the difference between these measures when each dog has been fed differently. If we want, we could even turn these measures into a difference score by subtracting the speed on the high-protein diet from the speed on the standard diet for each dog in turn. Doing this makes it immediately clear that what looks like two measures of each individual in our study can in fact be thought of as a single measure of change in a dog's performance. And if we only have a single measure per dog of the quantity we are interested in (i.e. difference in speed), we are not pseudoreplicating. Now, whilst it is true that the situation may become harder to visualize for more complex within-subject designs, as long as we analyse our experiment appropriately to take into account the within-subject measures, the same general principles apply. If the experiment had been done as a between-subjects design, with 10 of the dogs being fed the normal diet, and 10 being fed the high-protein diet, and then each dog had been measured twice, then the two measures on each dog would have been clear pseudoreplicates, and we should combine measurements to get a mean score for each dog before doing any statistical analysis.

 Repeatedly measuring the same individual within a well-planned within-subject design is not pseudoreplication.

10.5 With multiple treatments, within-subject experiments can take a long time

Let's consider the chickens and nestbox lining study a little further. Imagine that you are going to use four different treatments. The straightforward design would be to randomly allocate individuals to different groups, each of which experiences the four regimes in a different order. Now you can see a drawback, especially if we include wash-out periods—this experiment is going to take a very long time. There may be practical reasons why this is inconvenient. Further, the longer the experiment goes on, inevitably, the greater the problem drop-outs will be.

There may also be ethical considerations. This is unlikely to be a problem for your chickens, but if the experiment involves stressful manipulations of animals, then single manipulations on a larger number of animals may be a better option than repeated manipulations on the same subjects. This is a difficult decision to make; but it's a biological one, not a statistical one, and you cannot avoid it.

You can shorten the length of the experiment by using an incomplete design where each subject only experiences some (but not all) of the treatments. However, by now you will be getting the picture that we advise against incomplete designs if at all possible, because they are more challenging to analyse and interpret and certainly beyond the scope of this book. However, sometimes they will be unavoidable for practical reasons and in this case we recommend that you consult one of the texts in the bibliography and/or seek advice from a statistician.

 Within-subject designs with several treatments can take a long time to complete.

10.6 Which sequences should you use?

Until now we have made the henhouse study very small-scale by only looking at two treatments (hay and rubber matting). However, the experiment might be improved if we had a control treatment of the previously used substrate (wood shavings), to allow us to explore not just what alternative is better but also whether that alternative is better or worse than the one used previously by this farmer. This takes us up to three treatments; and if we want to explore a third alternative (shredded paper say), then this takes us up to four. With the four treatment groups of the henhouse experiment, there are 24 different orders in which a single henhouse experiences all four treatments. If you don't believe us, write them out for yourself. Until now, we have recommended counterbalancing where we make sure that all possible combinations are applied to at least one subject. This was not a problem when we had only two treatments, but with four treatments we would need at least 24 henhouses. If we had five treatments, that number would go up to 120, and up to 720 for six treatments. We can see that counterbalancing quickly becomes unfeasible if the number of treatments is anything other than small. In situations with a large number of treatments the normal approach is to randomize the order in which individuals experience the different treatments. That is, we select a random permutation of the treatments for each subject. We must select a random permutation independently for each subject. Each subject then experiences treatments in the order defined by their permutation. This will generally reduce the adverse impact of period effects but is not as robust as counterbalancing.

Both counterbalancing and randomization are methods of controlling for any period effects that may be present in your experiment. The advantage of counterbalancing is that your statistical analysis can test to see if order effects matter in your experiment as well as controlling for them. Randomization does not allow you to test for order effects in your statistical analysis, so you will not find out how strong they are, although the randomization methodology should help to avoid them confounding your experiment. As we mentioned in Section 10.3.2, if you are concerned about carry-over effects but determined to carry out a within-subject design then it is best to consult a statistician to help you decide on the best design.

➡ If the number of treatments is low (four or less) then counterbalancing sequences should generally be adopted; for larger number of factors, randomization is an attractive alternative.

10.7 Within-subject designs and randomized block designs

There is sometimes confusion between the randomized block designs discussed in Chapter 9 and the within-subject designs that we are looking at in this chapter. This confusion is not helped by the fact that, as we have seen, each kind of design is referred to by multiple names, and sometimes these names have been used interchangeably for the two designs. However, being able to discriminate between these designs matters because, when it comes to analysing the data, each design requires a different kind of analysis. So let's consider the two designs side by side. Once again we are interested in which kind of bedding we should use in our henhouses: hay or rubber matting. Both of our studies start with 10 henhouses, each with 10 hens living in them. For the first study, we proceed as above, and add rubber matting to five henhouses and then hay to the others. Then, after a period of time, we switch the type of bedding giving us a within-subject design. The second study is similar, except each henhouse contains separate enclosures for individual hens. We add rubber matting at random to half of the enclosures in each henhouse, and hay to the others, giving us a randomized block design (with each henhouse forming a block of ten nesting enclosures). These studies are, on the face of it, very similar. In each we have a number of henhouses and we apply both treatments to each henhouse. And in both studies our aim is to compare treatments within a given henhouse, reducing inherent variation caused by differences between our henhouses. The fundamental difference is that in the within-subject design the treatments are applied sequentially whilst in the randomized block design they are applied simultaneously. Why does this matter? Because things such as period effects and carry-over effects apply to within-subject designs, but not to randomized block designs; each leads to different statistical tests. If you think you have done a randomized block design, and choose the statistics appropriate for such a design, but in fact you have carried out a within-subject design, the conclusions you draw may be incorrect. If you are still struggling to see the distinction, Figure 10.1 compares these designs graphically.

 Within-subject designs and randomized block designs can appear similar, but it is important to be able to tell them apart.

10.8 It is possible to design experiments that mix within-subject and between-subject effects

The final design that we will discuss is one that mixes aspects of the previous chapters in a single study, including both treatments that are applied within-subject and other treatments applied between-subject. Suppose that you hypothesize that early nutrition will affect the performance of adult rats in one type of cognitive task, but not in another type of task. You will hopefully spot immediately that this is a question about an interaction between factors, and so we need to use a design that will allow us to test for an interaction (if you don't see this, go and read Chapter 8 again!) Your first thought would probably be to carry out a completely randomized two-factor study; and that would certainly allow you to answer the question. However, it might require a lot of rats, and we might be able to do better. An alternative is to do the following. First, randomly allocate rats to different treatment groups for the early nutrition factor (let's have two levels of this factor, and call them *poor* and *good*). Fairly obviously, a rat that has poor nutrition early in life cannot at some later point in life be switched to the other treatment group, so early nutrition status cannot be examined as a within-subject factor. However, an individual rat can probably be assessed in two different kinds of cognition task without serious welfare implications, and so we might decide that, rather than allocating rats from each nutrition treatment to one or other of the task treatment groups, we will instead test each rat in both tasks, one after the other. We would obviously need to keep in mind all of the issues that apply to within-subject designs, giving half of the rats one task first and the other rats the other task first, and allowing time between tasks to minimize carry-over effects. We also need to think very carefully about the statistics that we will use, as they will necessarily be more complex. Given this, why would we choose this design over the simpler, fully randomized design? Well there is one obvious advantage—by measuring each rat twice, once in each task, we have reduced the number of rats that we would need for a fully randomized study.

A similar type of design is often seen in medical studies where different medical procedures or drugs are applied to individual patients, and then some measurement is made on each patient at multiple time points, for example to monitor how the treatment affects the time course of their disease. So for example, patients might be given different flu treatments, and then their flu symptoms measured each day for one week. This is analogous to the previous design, with the subtle difference that our within-subject factor is not an experimental treatment like the bedding in the previous example. Instead, it is the day that we make our measurement. This can create additional complications when it comes to the analysis, since days of the week cannot be applied in a random order in the same way that our henhouse treatments could, but these need not bother us here (but we offer some suggestions for statistics books that cover this issue well in the Bibliography). However, our advice to plan your analysis at the same time as you plan your study becomes even more important than ever as your design becomes more complex. Before planning a study like this, make sure that it is the best way to address your question, and that you know how to analyse the data you will collect.

 More complex designs can include within- and between-subject factors, but more complex designs lead to more complex analyses.

Summary

- In a within-subject design experimental subjects experience the different experimental treatments sequentially, and comparisons are made on the same individual at different times, rather than between different individuals at the same time.

- Comparing within individuals removes many of the problems that between-group designs experience with noise.

- Care is required to avoid introducing time as a confounding factor, but this can be avoided with careful design.

- Within-subject designs are not suitable for situations where a subject cannot be returned to the condition it was in before the treatment started.

- Make sure that you leave sufficient time for a treatment to stop having an effect on subjects before applying the next treatment and taking measurements of the effect of the new treatment.

- Repeatedly measuring the same individual within a well-planned within-subject design is not pseudoreplication.
- Within-subject designs with several treatments can take a long time to complete.
- If the number of treatments is low (four or less) then counterbalancing sequences should generally be adopted; for a larger number of factors, randomization is an attractive alternative.
- In principle a design may include a mixture of within-subject and between-subject effects. Such designs can be very powerful for certain questions, but require careful analysis.

11

Taking measurements

If there is an over-riding message in this book, it is our championing of the need for careful thought and preparation before embarking on your main experiment. Chapter 2 covered several of the benefits of pilot studies. Another important benefit is that a pilot study gives you a chance to improve the quality of your recording of data. You will have spent a lot of time designing and carrying out your study, so you owe it to yourself to collect the best data that you possibly can. However, there are numerous ways that inaccuracy or bias can creep into your measurements if you are not careful. Here we will illustrate this with a number of examples and suggest ways in which common pitfalls can be avoided.

- We begin by illustrating the importance of checking the performance of measuring instruments through a calibration process (Section 11.1).

- We then discuss the dangers of imprecision and inaccuracy for continuously measured variables (Section 11.2), and subsampling as a way to control imprecision (Section 11.2.1).

- We discuss sensitivity and specificity as descriptors of the effectiveness of a diagnostic test of a dichotomous condition (Section 11.3).

- People are not machines, and their performance can change over time, so we discuss how to minimize the impact of this on your study (Section 11.4).

- Similarly, two people can record the same event differently, and so we discuss how to cope with this (Section 11.5).

- We move on to a section on some practical issues that can impinge on your data collection (Section 11.6). Specifically:

  - Thinking carefully about how you define levels of a categorical variable (Section 11.6.1);

  - Deciding how accurately you should measure continuous variables (Section 11.6.2);

- When human judgment is required in taking measurements, being on your guard against accusations of conscious or unconscious bias (Sections 11.6.3 and 11.6.4);
- When testing individuals, making sure the test is neither too easy nor too hard (Section 11.6.5);
- And making sure when dealing with live organisms that your attempt to measure does not influence the subjects in the trait that you want to measure (Section 11.6.6).
- We finish in Section 11.7 by pointing out some further data-collection pitfalls, caused just by human nature, that can be easily avoided with a little forethought.

11.1 Calibration

No matter how simple a measurement appears, there is always reason to be careful in order to avoid obtaining inaccurate results. Take the simplest possible case, where you are simply reading the mass of objects that are placed on a set of scales with a digital readout. Surely nothing can go wrong here? Well, do you know that the scales are accurate? Would it be possible to check by measuring some items of known mass? It is unlikely that your scales are faulty, but it takes little time to check. It would do no harm to do so at the end of your series of experimental measurements as well as at the beginning, just in case they started off working fine but something went wrong during your study. Imagine that your experiment involves carrying out a measurement once a day for 14 days. If you leave the scales on a laboratory bench, then it is not outside the bounds of possibility that someone could knock them off the bench then stealthily return them to the bench without telling you, or drop something heavy on them, or spill a can of soft drink on them. Hence, there is no harm in checking a measuring instrument by getting it to measure something for which you already know the correct value, a process called **calibration**. By taking a little extra time to incorporate this into your experiment, you will have improved confidence in your measurements.

Calibration involves checking or setting a measuring instrument by using it to measure items whose measurements are already known with certainty.

For our example of a set of scales, an alternative to measuring something of known mass would be to borrow a second set of scales for a short time and check that an object measured on the two sets of scales gives the same value in each case. Of course, both sets of scales could be faulty, but it is unlikely that both would be faulty in the same way (say, both underestimating by 7%), especially if the instruments are of different *types*. Also it is often worth checking that an instrument gives the correct value in some trivial case where the correct answer is obvious. For example, you could check that your scales read zero when nothing is placed on them.

You could reduce the problem of a fault developing during your set of measurements by saving your samples and measuring them immediately one after the other at the end of the 14 days. However, be careful here, as this means that some samples will be stored for longer than others. If weight is likely to change during storage, then this could introduce a confounding factor into your experiment. You should see devising a

Q 11.1 Imagine an experiment in which you aim to compare measures of the skeletal size of one-day-old chicks produced by captive zebra finches on two different diets. One of the diets appears to reduce the chick incubation period, hence causing the chicks born to mothers on that diet to be measured on average earlier in the experiment than those on the other diet. There seems no way to avoid time of measurement as a confounding factor, so what should you do?

You can find more material on the issues raised in this section in the equivalent section of the supplementary information. Go to: **www. oxfordtextbooks.co.uk/orc/ ruxton4e/**.

protocol for measuring as part of your experimental design and produce a design that best suits the biology of your system.

Of course, it goes without saying by now that randomization is important in measuring. Imagine in the previous example that we have seven samples from a control group and seven samples from a treatment group. These 14 samples should be measured in a random order (or possibly in an order that alternates control and treatment samples). This avoids time of measurement becoming a potential confounding factor. Why should this matter? First of all, it might be that the set of scales changes over time (perhaps as the room heats up during the day) or that the care with which you handle the samples and place them on the scales deteriorates (as it approaches lunchtime). By randomizing, you are not only removing this potential confounder, you are removing any perception that it might be important for your results. If these precautions seem excessive, then think of it in terms of satisfying the Devil's Advocate mentioned throughout this book. Also consider that these precautions generally involve very little extra effort.

→ Calibrate your instruments of measurement and avoid introducing time of measurement as a confounding factor.

11.2 Inaccuracy and imprecision

Imagine that you are measuring the times that it takes students to run 60 metres using a stopwatch with a traditional clockface rather than a digital readout. Each student has been subjected to one of two different types of fitness regime, and your experiment aims to explore whether these regimes differentially affect sprint speed.

The clockface introduces a greater need for interpretation by you than a digital readout does. You have some discretion in how you read off the time, since the second hand will not stop exactly on one of the 60 second-marks around the face. You must then either round up or down to the nearest second, or estimate what fraction of a second beyond the last full section the watch is indicating. Our first point is that a digital readout takes away this source of potential **imprecision**. However, if you must use an analogue watch, then make sure that you establish clear rules for how you estimate the time that you record from the watch. For example, you might decide to always round to the nearest even number of seconds.

This seems a good point to introduce the ideas of inaccuracy and imprecision. Your judgement in estimating a time from the stopwatch readout introduces a potential source of error. This means that there might potentially be some added noise in your data: if four students all took exactly 7.5 seconds to run 60 metres, you might record these not as {7.5, 7.5, 7.5, 7.5} but as {7.5, 8.0, 7.0, 7.5} just because of your variation in the way you operate the stopwatch and read the clockface. This imprecision adds between-individual variation to the results, but is random in the sense that it doesn't systematically lead to high or low readings. Imprecision is something that we should seek to minimize in our experimental design. Now what if your stopwatch actually runs fast, so that you actually record times of {9.5, 9.5, 9.5, 9.5}? This is **bias** or **inaccuracy**.

Imprecision adds errors to your measurements but in a way that is uncorrelated from one measurement to the next. That is, if one measurement is underestimated then the next measurement might be either under- or overestimated. **Bias** or **inaccuracy** also add errors, but in a correlated way. That is, if one measurement is underestimated by a certain amount, then the next measurement is likely to be similarly affected.

This is generally a more serious problem than imprecision. If all we are doing is trying to test whether the two groups of students differ in sprint speed then there is not too much of a problem, since the bias caused by the faulty watch will be the same for both groups. However, if we wanted to quantify any difference in sprint speed then we would have problems. Similarly, we have problems if we want to compare our results to those recorded with a different stopwatch that does not have the same bias. Worse still, there is no way that your statistical analysis will tell you if you have a bias problem; in fact it will assume that you definitely do not. Hence, you can see that there is a very real reason for calibration, as only this will detect bias. It is sometimes possible to correct your data to remove the effect of a bias. Say, in our current example, that we worked out how fast the stopwatch was running, then we could correct our recorded times appropriately to get the actual times. However, this can only occur if we can detect and quantify the bias. Calibration checks—like those suggested in Section 11.1—are the best way to achieve this. Figure 11.1 presents a classic way to illustrate inaccuracy and imprecision, with the results of several shots, all aimed at the bullseye of a target.

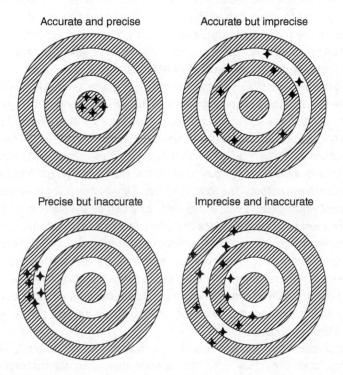

Figure 11.1 The four ways that precision and accuracy can interact, illustrated as the grouping of a number of shots aimed at the centre of the target. Inaccurate shooting will show a consistent bias; in the example here we see that if the shooting is inaccurate then shots are consistently further to the left than the ideal. Imprecise shooting shows variation that is not consistent from shot to shot, so some shots are too high, other shots are too low; some shots are too far to the left, others too far to the right. Precise shooting will result in a closely spaced, tight grouping of shots, whereas imprecise shooting will show a wider spread.

There are other areas of our imaginary experiment where imprecision and bias can creep in. Imagine that you stand opposite the finishing line and start the stopwatch when you shout 'Go!'. There could be a tendency for you to press the stopwatch just fractionally before you shout. There is also an element of judgement in deciding when individuals cross the line. What if some students dip on the line like professional athletes and others do not? One way to limit this imprecision is to have a clear rule for when the person crosses the line. Something like, 'I stop the stopwatch when the first part of the person's torso is above the finish line'.

Another potential source of bias in this experiment is that you (standing at the finishing line) are taking it on trust that the people are starting at the start line. In order to reduce their time, individuals might encroach over the line a bit. Further, if individuals see previous students get away with this cheating, then each is likely to encroach a little more than the last. This could lead to completely false results if you have not randomized the order of measuring individuals. The moral here is to try and design experiments so that there is as little scope for inaccuracy and imprecision as possible. In the running experiment, you might do well to have a colleague checking that the students all start from just behind the line. You should not be worried that students will be insulted by the implied suggestion that they are potential cheats. As ever, you are satisfying the Devil's Advocate and removing the *perception* that this might be a potential source of problems in your measurements.

Allied to the last point, this is exactly the sort of experiment where it would be useful if you were 'blind to' the treatment groups that individuals belonged to, until after you have measured their run (we will discuss the idea of blinding fully in Section 11.6.3). The key issue here is concern that the person measuring running times might be biased by preconceptions about the relative effects of the two treatment regimes. That is, because they expect individuals in one group to run faster then (perhaps unconsciously) they tend to record them as faster than they really are. If the person doing the measurement is unaware of which group individual runners belong to then this bias cannot arise.

All of this must be making you think that a lot of these problems could be removed if you measured the time electronically, with the clock starting when one light beam (over the start line) is broken and ending when another (over the finishing line) is broken. This would certainly remove some potential sources of error: it should reduce inaccuracy and also concerns that a human who is measuring speeds on a stopwatch might be biased by their preconceptions. However, remember that it would still be worthwhile calibrating this automated system.

Since imprecision adds random noise to our measurements it will act to reduce the power of our study. In essence, the additional noise makes any patterns harder to see in our data. As we saw in Chapter 6, the most obvious way to increase the power of our study is to increase the number of replicates. Imprecision in our measurements of the running speeds of students will be caused by both your measurement technique and by the performance of the student (for example, they might slip slightly when starting, or be slow to start through some external distraction), and by measuring a larger number of students we more effectively average over these effects. However, sometimes this will not be feasible, for example if there are constraints on available equipment or biological samples. In other cases, especially if we are working with live animals, increasing replication may not

Q 11.2 Would you use the electronic system to measure running speeds in the previous experiment on human running speed?

be desirable for ethical or welfare reasons. In these situations, increasing the precision with which we measure each replicate by taking multiple measurements on each to average across sources of imprecision can provide an effective way to decrease inherent variation and to increase the power of our study. Thus, we might choose to measure each student running on several different occasions and use the mean of these runs as our measure of the characteristic running speed of that student. Notice that we would include only one value (the mean) for each student in our analysis, to avoid pseudoreplicating.

➡ Take time to demonstrate to yourself and to the Devil's Advocate that the biases and inaccuracies that exist in your experiment are small enough to have a negligible effect on your conclusions.

11.2.1 Subsampling: more woods or more trees?

In the previous section, we outlined two different sampling approaches that might be used to counteract imprecision, increase the total number of replicates or increase the precision of our measure of each replicate by measuring it several times. A problem that will crop up in many studies is how to allocate your sampling efforts between these two different levels. We will illustrate this with an example. Imagine you are interested in testing the hypothesis:

The numbers of species of beetle found on trees differ between local coniferous and broadleaf forests.

You consult your map and find ten possible forests of each type, and you decide that you have the resources and time to examine 100 trees in total. How should you allocate your sampling across forests? One solution would be to choose one forest of each type and then examine 50 trees in each of these forests (selected at random within forests, of course). This would give you the maximum amount of information on each forest, but on the minimum number of forests (only one of each type). If you are interested in general differences between coniferous and broadleaf forests this would clearly not be the way to proceed. Another alternative would be to sample all 20 forests, which would mean that you could examine only five trees at each site. This would give you much less information on each site, but would give you at least some information on the maximum number of forests. Alternatively, you could compromise between these extremes; and sample five forests of each type, examining ten trees in each. Which of the three designs (illustrated in Figure 11.2) is best?

To answer this question we need to go back and think about the purpose of replication. Replication allows us to deal with variation. In this example we have two sources of variation:

- Trees within a single forest will vary from each other in how many beetle species they carry.
- The forests will differ between each other in the average number of beetle species that the trees within them carry. This may occur because of differences between forests in altitude or hydrology, for example.

@ You can find more material on the issues raised in this section in the equivalent section of the supplementary information. Go to: **www.oxfordtextbooks.co.uk/orc/ruxton4e/**.

❓ **Q 11.3** In the text, we say that if you are interested in general differences between coniferous and broadleaf forests, then studying only one forest of each type is 'clearly not be the way to proceed'. Can you explain why?

Three designs for sampling 50 conifer trees

a)

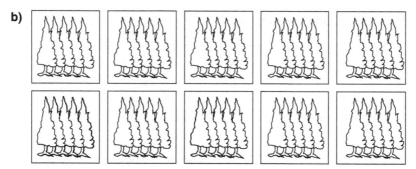

50 trees all from the same forest. **Excellent** information about that forest but no information on other conifer forests.

b)

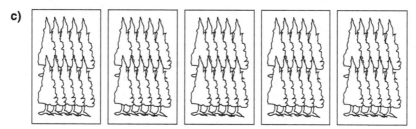

5 trees from each of 10 forests. **Fair** information of a good sample of different conifer forests.

c)

10 trees from each of 5 forests. **Good** information on a fair sample of different conifer forests.

Figure 11.2 We can sample 50 trees from forests of a particular type. Our independent subject is a forest. If the maximum number of forests of a given type available to us is ten, then one approach would be to utilize as many subjects as we can, and measure five trees from each of ten forests. For each forest we might report mean number of beetle species found on each tree. However, if trees within a forest are very variable then a mean of just five trees might be unrepresentative of trees in that forest. To combat this, we could increase the extent of subsampling and take an average over ten trees in each forest. The price that must be paid for this is that for the same sampling effort we can now only investigate half the number of subjects (five forests). In the most extreme case we could subsample very extensively and collect the best information we can on one forest of the particular type (sampling 50 trees in that forest). In general, our advice is to bias sampling so as to best control the source of greatest variation. In our example, if we expected trees within a forest to be much more variable in diversity of beetles than differences in average per-tree diversity between forests, then we should subsample more extensively at a cost of having fewer subjects. However, if in doubt, seek to maximize your number of independent subjects.

We need to bias our sampling towards the source of the greatest amount of variation. At one extreme, imagine all trees within a forest have exactly the same number of beetle species. That is, imagine that there is no variation between trees within a forest. In such a situation, we would only need to sample one tree within each forest to get a good estimate of the number of beetle species per tree at that site, and we could concentrate on measuring as many different sites as possible. At the other extreme, what if all the coniferous forests have the same average number of beetle species per tree, and all the broadleaf forests have the same average number as each other (but different from the conifers), but trees within a site vary (about the appropriate average) in the number of species they carry. In this case there would be little need to sample multiple forests of each type and we could concentrate on getting better data from a single example of each type of forest. Thus, you can see that the answer to the question depends on the biological variation that is present. With the benefit of pilot data, it would be possible to estimate the power of designs with various different sampling regimes to determine which is the best one to use. Such techniques are similar to those discussed for simple designs in Chapter 6, but this is not the place to discuss them further. The book by Underwood (1996) in the Bibliography has a particularly thorough consideration of these issues. *However, as a good rule of thumb, unless you believe that the variation within independent experimental subjects (forests in this case) will be very large, or you are already sampling many subjects, increasing the number of subjects will generally give you more power than increasing the number of subsamples (trees in this case) from a subject.*

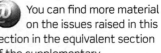 You can find more material on the issues raised in this section in the equivalent section of the supplementary information. Go to: **www.oxfordtextbooks.co.uk/orc/ruxton4e/.**

 Bias your sampling towards getting more measurements at the level where you expect the greatest variation.

11.3 **Sensitivity and specificity**

For some studies you might want to classify individuals into one of two categories on the basis of some underlying condition that is difficult to observe directly; for example, whether or not a person is in the early stages of pregnancy, or whether a particular badger carries tuberculosis. In these situations it is common to make use of some form of diagnostic test that classifies individuals into categories based on some indirect measure. Since we are not observing the condition directly, but instead relying on an indirect measure, there is an obvious question about the reliability of our classification. **Sensitivity** and **specificity** are summary statistics of the performance of such a diagnostic test (analogous to accuracy and precision for continuous measured traits). The sensitivity is the proportion of truly positive individuals that are recognized as positive by the test. Thus, if a urine test classifies 80% of truly pregnant women as pregnant then it has a sensitivity of 80%. That is, in 20% of cases where the test is applied to a truly positive individual, a negative result is erroneously reported. To put it another way, there is a 20% rate of false negatives. The specificity is the proportion of truly negative individuals that are classified as negative by the test. Thus if a test for pregnancy had

For a binary classification test, the **sensitivity** is the proportion of individuals that truly have the condition being tested for, and that are reported as having the condition by the test. **Specificity** is the proportion of individuals that do not have the condition and are reported as not having the condition by the test.

a specificity of 95% we would expect that 95% of women who took the test and really were not pregnant would be classified as not pregnant by the test. The rate of false positives then is 5%.

Ideally a test should have both a high sensitivity and a high specificity, but often changes to the test that increase sensitivity will reduce specificity and vice versa. In order to assess whether a car driver is likely to have his driving ability impaired by alcohol consumption, the amount of alcohol in the breath is measured. If this is over a certain threshold then the test is positive for alcohol-impaired driving ability. The level of alcohol we set for our threshold will have opposite effects on the sensitivity and specificity of our test. Thus, if we set a very low threshold we will correctly identify as positive nearly all individuals who really are positive (our test will be sensitive). However, a low threshold will also mean that we would get many false positive results from people whose driving was unimpaired (our tests will have low specificity). We can increase the specificity by increasing our threshold, but an inevitable consequence will be that some drivers whose ability is in fact impaired will now test negative, and our test has become less sensitive. As a society, we have had to agree on a threshold that we feel strikes the best balance between the two types of error. An important factor in this decision has to be the costs (in our example, both to the driver and to other individuals) of the two types of errors.

Let's say that a home pregnancy test kit has a sensitivity of 97% and a specificity of 90%. If you take the test and it comes back positive, how likely are you to be pregnant? A common error that people make in this situation is to assume that the answer is simply 90%, the specificity of the test. Their logic goes something like, 'If the specificity of the test is 90%, then the false positive rate is 10%. So the probability that my positive test result is false is 10%. Therefore there is a 90% chance that my positive result is true and I am pregnant'. Whilst this logic may seem convincing (and you would not be alone in thinking so—in a study of qualified gynaecologists over 80% made this kind of error), it is utterly wrong. The reason is that the probability that a positive result from the test is a true positive depends not just on the sensitivity and specificity of the test, but also on the underlying probability that you are pregnant in the first place.

The **positive predictive value** of a test for a condition is the fraction of individuals that test positive for the condition that actually have the condition. The **negative predictive value** is the fraction of those individuals that test negative for the condition that actually do not have the condition. The **prevalence** of a condition is the fraction of people tested who actually have the condition at the time of testing.

To convince you of this, let's consider two extreme scenarios. First, imagine that one of the (male) authors of this book takes the test and it comes back positive. Short of a medical miracle, you would be 100% sure that this positive is a false positive, despite the fact that the false positive rate of the test is only 10%. Now suppose instead that we give the test to a woman attending antenatal classes and again get a positive result. In this case it is almost 100% certain that the positive result is a true positive. So we can see clearly that the probability that a positive is a true positive is not simply the opposite of the false positive rate, but depends on the underlying **prevalence** of the condition in the population from which the individual being tested is drawn. To deal with this issue, statisticians have produced two further measures: the **positive predictive value** (or PPV), which is the proportion of individuals that are scored by the test as positive that are actually positive, and its equivalent for negative outcomes, the **negative predictive value** (NPV). These measures take into account both the sensitivity and specificity of the tests, but in addition how common the condition is in the population being tested. Box 11.1 shows how they can be calculated.

In situations where increasing sensitivity reduces specificity or vice versa, how can you decide which to maximize in your own study? As we discussed earlier in relation to alcohol and driving ability, sometimes if there are differences in the consequences of false positives and false negatives we might choose our sensitivity to minimize the more serious type of error. Consider the metal detector that you walk though at airport security. The function of this test is to detect weapons that someone might attempt to smuggle on board the aircraft. The metal detector is set so that it produces a high rate of false positives in order to minimize the risk of false negatives. If the metal detector falsely indicates that you could be carrying a weapon, then this mistake (a false positive) is quickly and cheaply corrected by means of a physical search of your person by a member of security staff. It should be obvious to you that the cost of the machine's failure to detect someone that is actually carrying a weapon (a false negative) is likely to be much, much higher. However, there are other situations in which we might simply want to minimize the total number of errors, irrespective of type. For example, imagine you want to estimate the level of tuberculosis in a population of badgers. In this situation, minimizing the total number of errors will provide the most precise estimate, and we have no reason to favour reducing one type of error over another. In this case the optimal balance between sensitivity and specificity that minimizes the total number of errors (both positive and negative) will depend on whether you are studying a rare or a common phenomenon. In general, if you are studying a rare phenomenon you will want your test to be as sensitive as possible to reduce false positives, whilst for common phenomena a specific test which reduces the level of false negatives will be better.

@ You can find more material on the issues raised in this section in the equivalent section of the supplementary information. Go to: **www. oxfordtextbooks.co.uk/orc/ ruxton4e/**.

BOX 11.1 Positive and negative predictive values

In the main text we discussed how it is easy to draw erroneous conclusions when faced with a positive result in a diagnostic test such as a home pregnancy test. In this box we will examine the correct way to deal with this issue by calculating the positive predictive value (PPV) of our pregnancy test.

$$PPV = \frac{\text{number of true positives}}{\text{number of true positives} + \text{number of false positives}}$$

In order to make further progress, we need to define the prevalence. This is the fraction of test subjects that are actually positive for the quality under test at the time of the test. For our example, this is the fraction of women taking the test who are actually pregnant. From our definitions it is easy to see that:

$$PPV = \frac{(\text{sensitivity}) \times (\text{prevalence})}{[(\text{sensitivity}) \times (\text{prevalence})] + [(1 - \text{specificity}) \times (1 - \text{prevalence})]}$$

For our hypothetical example, we know the sensitivity (0.97) and the specificity (0.90). It is hard to guess the prevalence, so let's try a range of plausible values: 10%, 30%, and 50%. These yield PPV values of 0.51, 0.81, and 0.91, respectively. The important take-home messages here are that sensitivity and PPV can be very different and PPV depends very much on prevalence.

There is an equivalent negative predictive value (NPV) which is the proportion of those individuals that are scored by the test as negative that are actually negative.

$$NPV = \frac{number\ of\ true\ negatives}{number\ of\ true\ negatives + number\ of\ false\ negatives}$$

$$NPV = \frac{(specificity) \times (1-prevalence)}{[(specificity) \times (1-prevalence)] + [(1-sensitivity) \times (prevalence)]}$$

 No diagnostic test based on indirect measures is perfect, so we need to be on our guard against both false positives and false negatives.

11.4 Intra-observer variability

11.4.1 Describing the problem

In our next thought experiment, for a biogeographical study of changing land-use patterns in Amazonia you are given access to 1000 satellite images of 100 m-square sections of Amazonian rainforest. You are asked to categorize the extent of human disturbance in each. You decide to categorize them according to the following scale:

- 0: no signs of any human disturbance
- 1: minimal signs of transitory human disturbance (e.g. trails, litter)
- 2: more extensive but transitory disturbance (e.g. larger trails, unmade road, camp site)
- 3: substantial but transitory disturbance (e.g. a tarmacked road, evidence of agriculture or forestry)
- 4: low-intensity human settlement (e.g. scattered housing, a small farmstead)
- 5: high-intensity human settlement (e.g. a town or village, a factory)

> **Observer drift** is systematic change in a measuring instrument or human observer over time, such that the measurement taken from an individual subject depends on when it was measured in a sequence of measurements as well as on its own intrinsic properties.

You work through the 1000 photos, writing the appropriate category (using the 0–5 scale) on the back of each photo. This takes you three hours. We would really be concerned about the quality of this data. Our main concern is **observer drift** (sometimes called *rater drift*). To put it another way, we are concerned that the score a photo gets is influenced by its position in the sequence. This might happen for any number of reasons. First of all, three hours is a very long time to be doing this sort of thing without a break. There is a real danger that you got tired towards the end and sometimes wrote '3' even though you were thinking '4'. Also you can subconsciously change the way you define categories over time. Imagine that you get the feeling after doing 50 or so, that you haven't given any scores of 1; there is a natural tendency to become more likely to decide that the next few borderline cases between categories 1 and 2 are scored as 1.

11.4.2 Tackling the problem

There are lots of things you can do to avoid problems due to observer drift. First of all, you can see that randomization of the order in which you measure is a good idea: if these 1000 photos belong to two geographical areas that you want to compare, you don't want to evaluate all of the photos from one area then all of the other. Now imagine a situation where you know that you'll be given another 1000 photos at some future date. How can you help ensure that your judgement will be the same for this second set as for the current one? One thing you could do is avoid having a long gap between scoring the two sets if you can. If you cannot, then you could do one (or preferably both) of the following:

1) Try to introduce objective descriptions of the categories or the dividing line between them. For example, a division rule might be 'If the extent of non-vegetated cover in a permanent settlement is greater than 30% then score as a 5, otherwise as a 4'. By reducing the element of subjective judgment involved, you should reduce **intra-observer variability**.

2) Keep a portfolio of examples of different categories. Particularly useful would be examples of cases that are close to the borderline of categories. Looking over these type specimens before starting each further set of measurements should help to avoid drift.

Intra-observer variability is simply imprecision or inaccuracy introduced by human error.

11.4.3 Repeatability

Observer drift is a real problem in lots of situations. However, it has one nice feature: it is often really easy to test to see whether you are suffering from it or not. Imagine that you get the second set of photographs two months later. Get yourself ready to start on the second set, perhaps by reading your notes from the first set, and looking at the portfolio of type specimens and your objective definitions. Then instead of starting on the second set, score a random selection of 50 from the first set (taking care not to look at the scores that each got the first time). Now compare your two scores for each of these 50. If they are identical, then it seems that you have little to worry about, and you can move on to the second set of photographs confident that whether a photograph was in the first or second set will not influence its score. However, if there are discrepancies, then the nature of these should indicate to you the aspects of your scoring that have changed. For example, you may be tending to see human habitation as more intensive now than before.

Statisticians call this procedure of measuring random samples several times and comparing their values a **repeatability study**, for the obvious reason that you are seeing how well you can repeat the measurements that you made the first time.

There are a number of formal ways that can be used to put a figure on how repeatable a measurement is. The most common of these is a 0–1 scale called a repeatability score (sometimes called an inter-class correlation coefficient; often termed 'r'). See, for example, Krebs (1999) in the Bibliography for details of how to calculate this. Quoting an r value close to 1 is a succinct way of reassuring others of the repeatability of your

A measurement is highly repeatable if you get the same score each time that you measure the same thing, and has low **repeatability** if your scores tend to be inconsistent. Notice that high repeatability suggests low imprecision but tells you nothing about bias. You can be consistent, but consistently wrong!

measurements. However, we would always recommend plotting or tabulating your repeated-measurement data for your own benefit, to see if there are any trends in where your repeatability is worse than in other situations (see Figure 11.3 for examples). If you can pinpoint such trends, then this may help you to improve your measurement technique in future similar studies. Of course, you can look at the repeatability of any measurement, not just category measurements like our satellite images. Even in the study discussed above with the digital balance, it might be worth measuring several samples twice, to ensure that the weight measurements are repeatable. Knowing that your measurements are repeatable will give you greater confidence in their quality.

So it is easy to test whether you have a problem with observer drift. However, what can you do if it does appear to be a problem? In the rainforest-photos study earlier, we would recommend going back to the original set, and picking out cases that are closer to the border between two categories. Have a look at the scores you gave them before. See if you get a feeling for how you were scoring. If you get this feeling, then go and do some other stuff for a while and completely forget about Amazonia. After this break, come back and take the 50-photo test again. Don't just look at photos that were borderline between low and high density human habitation; take another random sample and see if your agreement between old and new scores is now perfect or at least much improved. If so, you are probably ready to score the full second set; if not just keep repeating this procedure until you get yourself back into the same (scoring) frame of mind as you were in for the first set. The bad news is that sometimes this won't happen. Sometimes you will just not be able to understand how on Earth you gave the first set the scores you did. When this happens, you have little choice but to re-score the first set at the same time as you are doing the second. This means that your original three hours were for nothing: you can see that it pays to try and have as objective a scoring system as possible, to reduce the chance of this happening.

Q 11.4 A study looks at domestic cats on different diets. You are a veterinarian blind to the experimental groups that cats belong to, but must measure body condition by taking a body mass and measuring the skeletal size of the cat along their back from the top of the head to the base of the tail. Owners bring their cats in for measurement only once, and you perform this task on four different cats each week for 20 weeks. How can you defend yourself against possible accusations about drift?

You can find more material on the issues raised in this section in the equivalent section of the supplementary information, including consideration of the term *validity* in the context of repeatability. Go to: **www.oxfordtextbooks.co.uk/orc/ruxton4e/**.

Observer drift can easily occur, but it is easy to check if it has occurred, and if you detect it then remedial actions are available. But prevention is better than cure.

11.4.4 Remember, you can be consistent but still consistently wrong

The methods described in the last section of remeasuring and comparing allows you to spot whether you are consistent in your measurements, but it does not tell you whether you are consistently right or consistently wrong. Consider a situation where you have to look at 100 X-rays of horses' legs and identify whether or not each shows a fractured bone. If you score all of these, then rescore them all and find that your assessments agree, then this is encouraging; you are not drifting in your assessment. However, is your assessment right? Where there is a degree of subjectivity (and generally there is a considerable degree in looking at X-rays), then there is the danger that you have a consistent bias. Perhaps there is a certain type of shadow that you tend to consistently misidentify as a fracture, or a certain type of hairline fracture that you consistently miss. How can you detect this sort of bias? One approach would be to get someone to

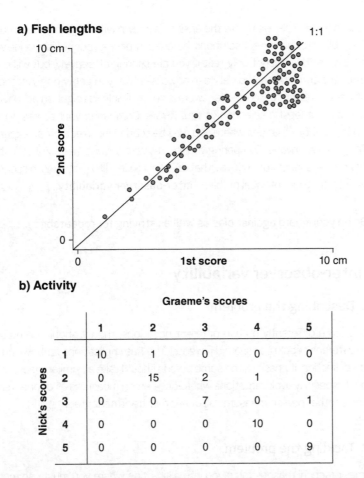

a) Fish lengths

b) Activity

Graeme's scores

	1	2	3	4	5
1	10	1	0	0	0
2	0	15	2	1	0
3	0	0	7	0	1
4	0	0	0	10	0
5	0	0	0	0	9

Nick's scores

Figure 11.3 Part (a) is a plot used to explore the repeatability of measurements of the length of some dead fish. Each individual fish in the sample was measured on two separate occasions, and each point on the plot has the first measurement as its x value and second as its y value. You can see that repeatability is very good for short fish, but for longer fish there are two concerning features of the data: (i) variability between the first and second scores (scatter away from the 1:1 line) is higher, and there is a tendency for first measurements to be higher than second measurements. This effect might be explained by drift in the method used to stretch out fish on the ruler that is only apparent in large fish. Alternatively or additionally, it may be a real physical effect with particularly large fish shrinking in length as they dry out. In part (b), Graeme and Nick both evaluated (separately) the extent of activity in a sample of separately housed woodlice using a previously agreed 1–5 scale. The tabulated results show good agreement between the two scorers (most woodlice are on the leading diagonal representing identical scores by Graeme and Nick). However, you can see a slight tendency for Graeme to give higher scores than Nick.

Inter-observer variability is error that arises because several observers were used to gather a body of data. Two observers (or more generally two different measuring instruments) are not exactly alike—differences between them can add imprecision and, if you are not careful in your design, bias.

 You can find more material on the issues raised in this section in the equivalent section of the supplementary information. Go to: **www.oxfordtextbooks.co.uk/orc/ruxton4e/**.

Q 11.5 What are the problems with different people measuring different treatment groups?

make you up a set of X-rays where the answer is known with certainty. You could then score these blind to the correct solutions, before comparing your performance with the correct answers. This will not only test if you are scoring effectively, but will also help you identify the common thread between any cases that you get wrong. Another alternative, if a set of absolutely correct answers is not available, is to get an acknowledged expert to score at least a selection of your X-rays. Comparing your scores with theirs may help to identify situations where you may be scoring incorrectly. If an expert is not available, then use anyone. Comparing your scores with anyone else's will be instructive in identifying the aspects of an X-ray that are likely to lead to problems in scoring. This leads us onto another potential problem: **inter-observer variability**.

➔ Be on your guard against bias as well as striving for repeatability.

11.5 Inter-observer variability

11.5.1 Describing the problem

If you can score differently on two different occasions, then it should be no surprise that two different people can score differently. This has implications for how you design your data collection. If possible, the same person should take all similar measurements. You should especially avoid situations where one person measures all of one treatment group and another person measures all of another treatment group.

11.5.2 Tackling the problem

Sometimes practical reasons force you into a situation where you must use more than one observer. The key thing here is to try and make as sure as possible that the two observers agree on how they score things. The way to explore this is exactly as we outlined in Section 11.3.3, but instead of you scoring the same set of subjects at two different times and comparing, now the different observers score the same set of subjects and compare their results.

Compared to intra-observer variability, it is generally easier to combat inter-observer variability, because when two of you disagree on scoring of the same thing, then you can have a discussion on why this has occurred and what you can do to prevent it happening again. As with intra-individual variability, reducing the need for subjective assessment is a sure way to reduce inter-individual variation. Similarly, don't just check to see if you agree with other recorders at the start of the experiment, check at the end too, to see if you have drifted apart. Finally, if you design your experiment correctly, and note down which observer made which measurements, you can even use the techniques that we described under blocking in Chapter 9 to deal with any small inconsistencies between observers (the same applies if you are forced to use more than one piece of measuring equipment in the same study, e.g. several stopwatches). In essence, we can treat each observer as an experimental block, and as long as each observer has taken measurements in all treatment groups (and we recommend very strongly that you make this the case to avoid problems of confounding), it is reasonably straightforward to control for differences between the observers in the same way as we did with differences between greenhouses. However, as with all things statistical, prevention is better (or at least easier to analyse) than cure, so avoiding inter-observer variability is worth aiming for.

→ Try to minimize the number of observers used, but if you must use several observers then adopt our recommended strategies to minimize, or allow you to control for, between-observer differences.

11.6 Deciding how to measure

In this section we want to briefly touch on some practical considerations that can improve your data collection: thinking carefully about how you define levels of a categorical variable (Section 11.6.1), deciding how accurately you should measure continuous variables (Section 11.6.2), when human judgment is required in taking measurements being on your guard against accusations of conscious or unconscious bias (Section 11.6.3), using allocation concealment as a defence against such biases (Section 11.6.4), when testing individuals making sure the test is neither too easy or too hard (Section 11.6.5), and making sure when dealing with live organisms that your attempt to measure does not influence the subjects in the trait that you want to measure (Section 11.6.6).

11.6.1 Defining categories

Assigning individuals to categories is often at the root of inter- and intra-observer variability. Take something as simple as watching rhinos in a zoo, and categorizing each individual's behaviour at 5-minute intervals as either 'standing', 'walking', or 'running'. Here are some things to consider.

First, we are glad that you have gone for three categories rather than 13. The more categories that you have, the more room for error there is. But you might consider a system where you record each rhino as either 'static' or 'moving' and if they are moving you record the rate of striding. This has the attraction that it gives you more precise information on movement rates and avoids having to make a tricky decision about when a fast walk becomes a run.

If you stick with the three categories, you have got to be sure that you can assign any given situation into one and only one of these categories. So you need to have a clear rule about how to differentiate between standing and walking and between walking and running. Is shifting of the feet whilst turning on the same spot walking or standing? Ideally all the observers should get together during a pilot study to work this out. They should finish with a clear statement for how categories are defined and how data are to be recorded, then they should try independently observing the same behaviour and see how they agree.

You can find more material on the issues raised in this section in the equivalent section of the supplementary information. Go to: **www.oxfordtextbooks.co.uk/orc/ruxton4e/**.

You can find more material on the issues raised in this section in the equivalent section of the supplementary information. Go to: **www.oxfordtextbooks.co.uk/orc/ruxton4e/**.

Define your categories as explicitly as you can, and test out your categorization scheme in a pilot study.

11.6.2 How fine-scale to make measurements of continuous variables

Whenever we measure on a continuous scale, for example, measuring the length of a snail's shell with a ruler, we need to make a decision about how fine-scale to make our measurements. Our ruler is marked in clear centimetre gradations so we could simply measure to the nearest centimetre. However, when we look closely we see that each centimetre on the ruler is broken up into millimetres, allowing us to make finer-scale measurements. Or we could go and get a pair of measuring callipers that measure to the nearest 0.01 centimetre. Which should we choose? The trade-off here is that the more fine-scale your measurements the more power you have at your disposal to detect and quantify effects of interest, but the more time, effort, and expense will likely be involved.

Imagine you are trialling two different food types for salmon under farmed conditions. At the end of your feeding regimes, when the fish are being harvested, you need to sample fish from cages subject to different food types and weigh each fish. You could use a spring balance. These cost £20 and measure the fish to the nearest 50 g. Or you could use a field-portable balance that costs £500 and gives you the mass to the nearest 0.5 g. Or you could take the fish back to the university to use the laboratory balance that costs £4000 and will give measurements to the nearest 0.01 g. This balance must be cleaned and recalibrated between measurements, a task that takes 30 minutes each time.

As with power analysis, in deciding an appropriate scale of measurement, you should have in mind the likely effect size of interest. In this case, no one would be very interested in a change in diet that yielded less than 1 g or 2 g extra to salmon that are generally around 2 kg when harvested. Hence, we suggest that use of the finest-scale balance would be too much work and expense for no substantial benefit over the medium-scale option. If we compare the remaining two options, if we feel that effect sizes of the order of 50–100 g would be of interest, then the cheapest method seems a little too crude in the resolution of data it would give us. The medium-range scale is not hugely expensive (compared to running a farm-scale feeding trial) and is probably not greatly different from the other option in terms of ease of use, portability, and reliability. Hence we would opt for the middle option as giving the best trade-off between statistical power and practical considerations.

 Embrace the best technology and methodology where it can yield valuable improvements in your statistical power; but explore whether simpler systems might not offer a more effective trade-off between statistical and practical considerations (such as availability, expense, time investment, and mechanical reliability).

11.6.3 Observer bias and blinding

You are always looking to convince the Devil's Advocate that there is no way that your recording of data can be susceptible to conscious or unconscious bias. Bias due to human involvement in measurement is called *observer bias* (or sometimes *experimenter bias* or *research bias*). Imagine you are trialling some antibiotics in the lab. You need to be able to defend yourself against the accusation that you find less growth of your pathogen on plates treated with an antibiotic because that is what you expect to see. The suggestion is that, when there is ambiguity in whether the pathogen has grown on a plate, your decision to record that plate as positive or negative for the pathogen might be biased by your expectation (this effect is often called *affirmation bias*). You can tackle this charge in two ways: by removing the ambiguity and by removing the expectation. Often a combination of the two is most effective.

You should design your data collection so as to have as little requirement for subjective judgement by the measurer as possible. Hence, instead of having someone simply look at the plates and pronounce whether each is positive or negative for the pathogen you should decide on some objective measure before measurement begins. In this case we might take digital photographs of the plates, decide on the colour range that codes for the pathogen, measure how much of the plate shows this colour using image analysis software, and have a rule for the minimum percentage that signals pathogen growth. This seems like quite a lot of work when you feel that ambiguous plates will occur relatively rarely. We agree: in this case it might be better to tackle the issue of expectation. This is particularly so since the approach involving photos can reduce but not necessarily remove subjectivity. For example, some photographs might come out poorly because you do not have the lighting regime just right, and you would naturally retake any such photos. But can you be sure that your decision to retake a photo is entirely free from bias relating to whether you expect an image of a given plate to show pathogen growth or not? All the more reason to remove expectation.

If the person who determines the pathogen status of each plate does not know which plates have the antibiotic, or indeed that there are even two different experimental groups (antibiotic-treated and untreated control), then they cannot be biased by an expectation that they do not have. This is a *blind procedure*. Our advice would be to design your data collection techniques to minimize requirements for human subjective judgment. Where appreciable levels of human judgment are required, then blind procedures should be implemented whenever possible.

Blind procedures are a powerful tool for silencing the Devil's Advocate, so let us look at them more closely. Recall previously that we discussed a clinical trial of the relative effectiveness of two wart treatments—let us imagine that they are both medicated creams—where we need to make some assessment of when a subject's warts have been cleared up. Now it would obviously make no sense if we use different criteria for the two different creams. However, what if the person that is making the assessment has some preconceived feeling about whether the new cream is likely to be effective or not? If this person also knows which people got which cream, then their prejudice may bias their assessment of when the warts are gone from a patient. Sometimes this might be deliberate deception on behalf of the researcher, although this is not usually the case. However, even if the researcher is not being deliberately dishonest, there is still the possibility that they will unconsciously bias assessments in line with their expectations. The obvious solution is to organize your experiment so that the

Q 11.6 Sometimes you cannot avoid subjectivity and cannot practically use blind procedures. Imagine that you are required to decide on the sanity of convicted murderers in a study that looks at gender differences. There is no objective test for sanity, so you are going to have to rely on the (at least potentially) subjective opinions of psychiatrists. These psychiatrists are probably going to have to interview the subjects, making it very difficult (if not impossible) for them to be blind to the subjects' genders. How can you defend the study against accusations of biased measurement?

A **blind procedure** is one in which the person measuring experimental subjects has no knowledge of which experimental manipulation each subject has experienced or which treatment group they belong to. In experiments with humans, we may use a **double-blind procedure** in which the experimental subjects too are kept ignorant of which treatment group they belong to.

person who judges the improvement in the patient as a result of treatment does not know which of the two treatments a particular patient received, or even that there actually are two different groups of patients. This is called a **blind procedure**, since the assessor is blind to the treatment group that an individual belongs to. The attraction of blind procedures is that they remove concern that the assessor may consciously or subconsciously bias their assessment.

In a blind study such as this, it is critical that, although the assessor is 'blind', someone knows which treatment group the patients belong to, or else the data collected wouldn't be any use. Such restricted access to information is often achieved by using codes. For example, each patient in the previous wart-cream example can be given an individual code number that can be entered on their patient record and accessed without restriction, provided someone (who should preferably not be involved with the assigning of patients to treatment groups, the treatment of the patients, or their final assessment) has a list that identifies which treatment group an individual code number refers to.

You should not worry that your collaborators will think that you are accusing them of being fraudsters if you suggest that blind procedures be used. These procedures are designed to remove the *perception* that *unconscious* bias *might* taint the results of a study. The Devil's Advocate will happily thrive on such a perception, and no one can reasonably claim that it would be impossible for them to have an unconscious bias.

Blind procedures are a particularly good idea whenever the experimenter is required to make any subjective measurements on subjects, although even something as apparently objective as timing with a stopwatch can be affected by observer bias (see Section 11.2). When the subjects are humans, then it may also be preferable to go one step further and ensure that the subjects themselves are blind to the treatment group that they are in. Such a procedure is called **double-blind**. To see the advantage of such a precaution, return to the wart treatments. Imagine that recent newspaper publicity has cast doubt on the effectiveness of the established method of treatment. It might be that patients assigned this treatment would be less motivated to apply the cream regularly and carefully, compared to those assigned to a new treatment that has escaped any negative publicity. This difference between the two groups could falsely inflate the apparent effectiveness of the novel treatment.

In order to facilitate blind and double-blind experiments, human patients are sometimes given **placebos**.

A **placebo**, or *vehicle control*, is a treatment that is designed to appear exactly like the real treatment except for the parameter under investigation.

For example, a placebo tablet would be made identically to the real tablet, except that the active ingredient under study would not be added. Notice that care should be taken so that the other ingredients are identical to the real tablet, and that the tablet and its packaging look the same (although a code would be used to identify it as placebo or real to the person finally analysing the experiment). Blind procedures involve a bit of extra work: but if your experiments involve humans as subjects or measurement tools then they are strongly recommended.

You can find more material on the issues raised in this section in the equivalent section of the supplementary information. Go to: **www. oxfordtextbooks.co.uk/orc/ ruxton4e/**.

Always ask yourself whether the Devil's Advocate would see any potential for biased measurements in your study: try to restrict their ammunition as much as you can.

11.6.4 Allocation concealment

Allocation concealment simply means that the allocation of particular subjects to particular treatment groups in an experiment is not known to anyone prior to subjects being allocated to particular groups. Imagine a situation where we are trialling a new treatment for a medical condition versus an established treatment, so we have two possible treatments that a subject could be randomized to if they take part in the trial. It is a good idea if the person is recruited into the trial prior to finding out which treatment they will receive, and that the person admitting them has no means of influencing which treatment they receive. For example, imagine a single doctor is going to be responsible for recruiting 50 individuals into the study. It would be useful if someone else generated the random allocation before recruitment begins by putting 25 'established treatment' cards in a bag and 25 'novel treatment' cards. These cards are drawn and each placed in separate opaque envelopes labelled 1 to 50. Only after the doctor has recruited the 27th person into the trial does the person responsible for allocating treatments to subjects open the envelope marked 27 to discover which treatment that person should be given.

Why we go through this elaborate procedure is to avoid conscious or unconscious biases impacting on the allocation strategy. Such *allocation bias* could occur if for example the doctor involved in recruitment had little faith in the novel treatment so as an act of kindness tended to allocate those individuals with the severest cases to the established treatment. If subjects were concerned about the safety of the novel treatment then they might be more willing to take part in the trial if they knew that they would be allocated the conventional treatment (this is *selection bias*), or might seek to influence the allocation procedure so that they were allocated to their preferred group (allocation bias). We can always perform allocation concealment and we always should in order to satisfy the Devil's Advocate that we have allowed no human biases to creep into our allocation strategy.

Notice that even after the person responsible for allocating the treatment to subjects has found out which particular treatment a patient has been allocated to there may be benefits to keeping this information from the patient themselves and from individuals involved in their care and assessment, again to avoid conscious or unconscious biases affecting results. Such withholding of information is called blinding and was discussed in Section 11.6.3. Blinding is sometimes not possible (e.g. a patient will know for certain whether they have been asked to perform a range of exercises or not), but allocation concealment always is, and so should always be undertaken. Allocation concealment and blinding are both strategies to avoid concerns about conscious or unconscious biases, and they are often confused. However, allocation concealment is about avoiding biases up to the point where treatments are imposed on experimental subjects, and blinding is about avoiding biases after this point.

For situations where individuals are recruited sequentially the most obvious thing to do would be to allocate them across the groups sequentially, so that with two groups the first person is allocated to group A, the second to group B, the third to group A, and so on. We do not recommend such non-random approaches because it is easy for individuals involved in allocation to predict the sequence and the Devil's Advocate will start to imagine ways that conscious or unconscious bias will start to creep in. There is no reason why simple randomized allocation cannot be implemented in sequentially recruited individuals, so it is much safer to do that.

➡ **Always be aware of the need to convince the Devil's Advocate that there are no biases in your allocation procedures. Allocation concealment is a useful weapon in this.**

11.6.5 Floor and ceiling effects

Imagine that you want to investigate the effect of alcohol on people's ability to absorb and recall information. Your design has two groups of ten people; individuals in the alcohol group are allowed to relax and listen to music for a couple of hours while they consume a fixed number of alcoholic drinks. Individuals in the control group experience the same regime except that their drinks are non-alcoholic. Each individual then watches the same five-minute film before being given a number of questions to answer about the film. The difficult part of designing this experiment is in preparing appropriate questions: they must be challenging but not too challenging. If the questions are so hard that none of the individuals in either group can get them correct then you learn nothing about differences between the groups. Similarly, if all twenty people can answer the questions without difficulty, then again you can learn nothing about the cognitive effects of alcohol. The first of these situations is a **floor effect**, in the jargon, the second is a **ceiling effect**. Ideally you want questions that are challenging but not impossible for both sober and intoxicated people to answer, so that you can measure differences in the extent of their accuracy in answering all the questions. The best way to avoid floor and ceiling effects is a pilot study.

A floor effect (sometimes called a *basement effect*) occurs when all (or nearly all) measurements taken from subjects are at the bottom value of the possible range. A ceiling effect occurs when all (or almost all) measurements on subjects are at the top value of the possible range. You should aim to avoid these effects, since you need to record a range of scores between individual subjects in order to have any ability to detect differences due to experimental treatments.

For example, say we randomly assign undergraduate volunteers to two different fitness training regimes. We wish to measure some aspect of fitness that we can then use to assess if one regime is more effective than the other. If we use the simple measurement—has 50-metre sprint speed improved after one month's training?—then we are likely to find that all individuals on both regimes can sprint faster after one month's training, and so we learn nothing about any differences there might be between the regimes. We have fallen foul of a ceiling effect. If, alternatively, we ask how many months on the training regime are required to reduce sprint speeds by 50%, then we run the opposite danger of setting a task that no one on either regime can achieve. We fall foul of a floor effect. Instead, we should seek to take a measurement where we expect that individuals will naturally be spread across a range of different values.

A **floor effect** occurs if the majority of measurements taken from subjects are at the lowest value of the possible range; a **ceiling effect** occurs when the majority take the highest value.

Perhaps we could measure the percentage reduction in 50-metre sprint speed after one month's training. We can then use a statistical test to ask how much (if any) of the between-individual variation in this measurement can be explained by the differences in training regime.

➡️ Avoid tests of subjects being too easy or too difficult for variation between individuals or groups to be picked up.

 You can find more material on the issues raised in this section in the equivalent section of the supplementary information. Go to: **www. oxfordtextbooks.co.uk/orc/ ruxton4e/**.

11.6.6 Observer effects

There are times when the simple act of observing a biological system will change the way it behaves. This is a phenomenon with a lot of names: you might find it called an *observer effect, observer-expectancy effect, experimenter-expectancy effect, expectancy bias*, or *experimenter effect*. Imagine you are watching the foraging behaviour of fish in a tank, both in the presence and absence of a plastic model of a predatory fish. Unfortunately, although the fish are not particularly frightened by your model predator, the fact that you are sitting over the tank watching them does worry them. As a consequence, all the fish behave as if there was a predator present all the time, and you cannot detect any difference due to the model's presence. These problems are not just limited to behavioural studies either. Suppose you want to know whether rodent malaria induces the same types of fever cycles as seen in humans. The simplest way to explore this would be to take mice infected with malaria and measure their temperature at regular intervals and compare this to control mice. However, the mere act of handling mice to take temperature readings may be enough to induce hormonal changes that increase the body temperature of the mice. The process of taking the measurement could make it an unreliable measurement. There are ethical and practical reasons why we would want to avoid the stress induced by testing our subjects in conditions that seem alien to them: we consider this in more detail in Box 11.2.

Phenomena like those described in the last paragraph present real problems to carrying out biological studies. How can we deal with them? One way is to allow the animals to acclimatize. Whenever we take animals into the lab, it is essential to give them time to settle, so that their behaviour and physiology are normal when we carry out our studies. One important part of the acclimatization might be getting used to the observer, or the handling procedures being used. Often observer effects will disappear after a suitable time of acclimatization. If this doesn't work, a next course of action might be to try to take measurements remotely. Our fish in the tank could be watched from behind a screen, or from a video camera. Similarly, temperature loggers are available that can be placed on the mouse, and will record its temperature at predefined intervals, removing the need to handle the animals during the experiment. Such equipment might be expensive, but might be the only way to get good data. It may also provide an ethical benefit if it reduces the stress that the animal suffers.

The biggest problem with observer effects is that by their very nature, it is hard to tell whether you have them or not. How do you know how the fish would be behaving if you were not there watching them? The answer is to try to take at least some measurement

> BOX 11.2 **Taking measurements of humans and animals in the laboratory**
>
> For both ethical and practical reasons you must avoid unintentionally stressing subjects. A stressed individual may not produce the same measurements of the variables that you are interested in as an unstressed individual in its normal surroundings. Stress is often caused by unfamiliarity with the surroundings. Build familiarization into your measurement procedures, such that you explain as much as is practical to human subjects before you start measuring them. Give them a chance to familiarize themselves with any equipment if this is possible, and encourage them to ask questions about anything that's on their mind. For measurements on laboratory animals, subjects are commonly moved from their normal accommodation to a test chamber. Both the process of moving and the experience of new surroundings are likely to stress the animal. Solutions to this are going to be highly species-specific but our advice is to study the experimental protocols of related previous works with a view to minimizing animal suffering and minimizing the danger of getting artefactual measurements because of unintentional stress caused to subjects. In summary, you can be sure that the potential for such abnormal behaviour will be something that occurs to the Devil's Advocate: do all that you can to prepare a defence against such concerns.

Q 11.7 One case where observer effects are very possible is when collecting data on humans by interview. How can you best address this?

 You can find more material on the issues raised in this section in the equivalent section of the supplementary information. Go to: **www. oxfordtextbooks.co.uk/orc/ ruxton4e/**.

remotely and compare these to direct measurements. It might be that you don't have access to enough video cameras to video all of your fish, but if you can borrow one for a few days and carry out a pilot study to compare the way fish behave with and without a human observer present, then you might be able to show convincingly that you don't need to video the fish in the main study. In the end, the question of observer effects will come down to the biology of your system, but you should do all that you can to minimize them, and then make it clear what you have done, so that others can evaluate your efforts. Always ask yourself whether the Devil's Advocate could claim that your process of measurement somehow affects the thing that you are measuring, and look to defend yourself against such a claim.

➡ Make sure that the way you collect information does not affect the variables that you are trying to measure.

11.7 Pitfalls to avoid when recording data

Recording data well is an art form. Doing it successfully will come in part with experience, and there are a number of tricks and tips you will pick up along the way. However, to get you started, here are some of the pitfalls that we wished someone had warned us about when we started. Some of these may sound trite to you; all we can say about the material in the following sections is that it is advice that we have ignored in the past to our cost.

11.7.1 Don't try to record too much information at once

This is particularly likely to happen in behavioural studies. You are only human, and the more different types of data that you try and record simultaneously the more mistakes you will make. Your pilot study will give you a chance to work out shorthand codes and data-sheet designs for efficient recording of information. If you really need to record a diverse stream of information from a behavioural study, then videotape it, and then watch the video several times, each time recording a different aspect of the data.

11.7.2 Beware of shorthand codes

Whilst watching a group of chimpanzees, you develop a code such that when an activity starts you say a two-digit code into a continuously running tape-recorder. The first digit identifies the individual chimp concerned (A, B, C, ... H) and the second is the type of behaviour:

- E: eating
- A: aggression
- P: play
- G: grooming
- V: vocalizing
- W: watching for predators.

Our first point is that it will be much easier to differentiate between 'Delta' and 'Bravo' than between 'b' and 'd', when you come to transcribing your tapes. Make sure that when you come to transcribing the tapes you can remember what 'W' was. Probably the safest thing is to define the code at the start of each tape that you use; that way the encoded data and the key to the code cannot become separated. The same goes for information on the date, location, weather, and whatever other information about the data collection you need to record—generally strive to keep related information together.

11.7.3 Keep more than one copy of your data

Data are hard and sometimes impossible to replace, but computers break down and notebooks get left on trains. Transcribe your fieldwork journal or lab-book onto a computer spreadsheet as soon as possible, keep a backup of the spreadsheet, and keep the original notebook. If it is impractical to transcribe it quickly, then photocopy pages regularly and post the photocopies to your mum, or at least keep them far away from your original notebook. If your house gets flooded we don't want both your original notebooks and your photocopy backups to be destroyed, that's why the copy is at your mum's. This may sound harsh, but if you lose data that you don't have a copy of, that's not being unlucky—it's having your carelessness exposed.

11.7.4 Write out your experimental protocol formally and in detail, and keep a detailed field journal or lab book

Imagine that you have just finished an intensive three-month field season watching chimpanzees in the wild. You deserve a holiday. After taking three weeks off to explore Africa, you come back home. It takes you another couple of weeks to work through the backlog of mail that has accumulated while you were in Africa. Hence, by the time you come to collate and analyse your data, it is 1–4 months since you actually collected those data. Expect to remember nothing! Everything to allow you the fullest possible analysis of your data must be recorded in some form (paper, computer disk, or audio tape). You cannot trust your own memory. Record everything that matters, and record it in a full, clear, and intelligible form. The test you should use is that your notes should be sufficiently full that someone else with a similar training to you (who has spent the last three months looking after your cat whilst you were in Africa) could take your data and analyse it as well as you could. So don't record the location as 'near the top of the big hill', give the GPS coordinates, or mark the spot on a map. Your notes should also include detailed descriptions of the categories you have been measuring, and any other measurement rules you came up with (did you measure arm length with the fingers included or not?). Careful note-keeping also has the added advantages that it makes it far easier for someone else to collect data for you should you become ill for a few days, and allows someone to follow up your research easily in the future.

11.7.5 Don't overwork

A corollary of the last point is that there is a tendency to feel that you only have three months to watch these chimpanzees, and so every waking minute not spent watching them is a minute wasted. Striving to collect as much data as possible is a mistake. Quantity is comforting, but this feeling deceives you: quantity is no substitute for quality. We will not mount our hobbyhorse about pilot studies again. However, time spent transcribing, collating, and checking the quality of your data is also time well spent, and will let you spot problems as soon as possible. Also, the sooner you transcribe your field notes into a more ordered form (probably by typing them into a computer) the less time there is for them to get lost or for you to forget important pieces of information. Most difficult of all, you must take breaks and do things unrelated to your work (sleeping, eating, and bathing are all to be recommended). This is not about admitting that you are weak or not sufficiently committed to your research, it's about believing that you record data better when you are fresh than when you are worn out. Don't be swept along by a macho culture that says you have got to suffer to collect good data. Tired recorders make mistakes and they don't notice when they are doing something stupid. Working too hard can actually be counterproductive and is potentially dangerous: don't do it!

Q 11.8 How long can you look down a microscope and count pollen before your accuracy begins to deteriorate through fatigue?

11.7.6 Take care to check computers and automated data collection

Computers are a real boon in data collection. Compared to humans, they are generally more precise, less likely to drift, and less likely to get tired or bored. However, their one big drawback compared to humans is a lack of self-criticism: they will record rubbish (or nothing) for days on end without giving you any indication that something is wrong. Hence it is essential that you check that your computer is recording what you think it should be. Pilot studies provide an ideal opportunity for this. Keep rechecking during the course of the study.

 Above all else: never, never, never have only one copy of a set of data.

■ Summary

- Your design must include consideration of how you'll take measurements.
- Calibrate your measuring instruments (including human observers).
- Randomize measurements.
- Consider subsampling to improve precision.
- Strive to demonstrate that the biases and inaccuracies that exist in your experiment are small enough to have a negligible effect on your conclusions.
- No diagnostic test based on indirect measures is perfect, so we need to be on our guard against both false positives and false negatives.
- Watch out for observer drift, intra-observer variability, and inter-observer variability.
- Adopt clear definitions to reduce subjective decision-making during measurement taking.
- Watch out for observer effects, where measuring a system influences its behaviour.
- Beware of floor and ceiling effects.
- Be ready to defend your study against accusations of observer bias and/or stressed live subjects.
- Always back up your data.

Sample answers to self-test questions

Remember that these answers are not the only possible answers to the relevant questions. Many of the questions are deliberately open-ended, and your answer might be quite different from ours but equally valid.

Chapter 1

Q 1.1

The question is rather vague as to the correct population to use. In particular, it mentions no geographical restriction. It is unlikely that the questioners are interested in all the prison inmates in the world. Hence, for convenience, we will assume that the questioners intend that we limit ourselves to the UK, but you could equally as validly assume any other geographical restriction. Since the prison population is also fluid, we also need a temporal restriction in order to exactly specify the population. Hence, if we decide to do our sampling on 12 March 2016, then the population from which we draw samples might most accurately be specified as 'all people being held in UK prisons on 12/3/16'. We should always look to be as specific as possible when describing our population, something that the original question did not achieve very well.

Q 1.2

For religious practices they are certainly not independent. Your expectation is that being locked in prison is a life-changing experience, where someone's openness to new religious ideas might be unusually high. Being locked in the same small room as someone who has strong religious feelings is highly likely to have a strong effect on someone's own religious feelings. Left-handedness is likely to be much less open to such social influences, so it might be reasonable to expect that the two cell-mates could be treated as independent samples for this part of the study. As with so much of experimental design, you'll see that arguments about statistical-sounding concepts like 'independence of data points' are addressed by our knowledge of the system rather than by mathematical theory.

Q 1.3

Take your pick from hundreds! We'd certainly expect age, gender, genetics, occupation, diet, and socioeconomic group all to have the potential to be linked to quality of eyesight. The problem is that these factors are also likely to be linked with someone's propensity to smoke, so teasing out the effects of smoking from other factors will be challenging.

Q 1.4

You could recruit a colleague to take measurements of one flock at the same agreed time as you are taking measurements on the other. However, this simply swaps one confounding factor (time of measurement) for another (person doing the measuring). Perhaps this would be an improvement if you could take steps to make sure that you both measure sheep in exactly the same way. Such steps could include both agreeing to a very explicit fixed protocol for sheep measurement and using the same type of equipment as each other. You could perhaps take half the measurements on one flock in the morning, then swap over with your colleague to take half the measurements on the other flock in the afternoon. You could then statistically test to see if the samples taken by the two different people from the same flock are significantly different from each other: if they are not, then this would give you confidence that you do not seem to have problems with identity of measurer or time of measurement as confounding factors in your study. However, there are lots of alternative ways you could have approached this study; this is by no means the only solution.

Chapter 2

Q 2.1

It may be that the people are (at least subconsciously) anxious to begin tackling the day's tasks at work as quickly as possible, whereas they are more relaxed about time on the way home. It could be that the penalty for arriving late at work is greater than the penalty for arriving late home from work. It could be that people like to lie in their bed as long as possible in the morning, then drive quickly to make up the time. It could be that the roads are generally busier when people return from work and so people are more likely to be delayed by slow-moving traffic on their return from work.

Q 2.2

It could be that regularly playing computer games does increase individuals' propensity to violence. One prediction stemming from this hypothesis is that if we found some volunteers who had not previously routinely played computer games and tested their propensity to perform violent acts both before and after a period of regular games playing, then their violence rating should increase.

   Another hypothesis is that computer games do not influence propensity to violence, but people who have a high propensity to violence for other reasons (be they genetic, environmental, or both) happen also to enjoy playing computer games. If this hypothesis is true, then we would expect that if we performed the experiment described above, then the volunteers' violence ratings would be unchanged by their period of regular games playing. Thus the experiment would allow us to test between our two rival hypotheses.

Q 2.3

Asking someone 'please keep a note of how often you commit violent acts over the next 6 months' or 'how often have you committed violent acts in the last 6 months?' is unlikely to produce very reliable answers. Firstly, people will disagree on what constitutes a violent act. Even if this can be overcome by agreeing a definition, then this is the sort of question about which you have to expect people to often be less than honest in their replies. We could stage situations involving potential conflict and monitor the reactions of our subjects. However, as well as being logistically very challenging, this is ethically highly questionable. Further, it is not clear to what extent someone's behaviour in one staged situation acts as a good description of their general propensity towards violent behaviour.

See Section 2.2.1 of supplementary material for more on the special challenges of dealing with people as the subjects in your study. Go to: **www.oxfordtextbooks. co.uk/orc/ruxton4e/**.

Q 2.4

Samples in the control group should differ from the samples in the experimental group only in the variable or variables that are of interest to the experiment. This scientist has inadvertently added a confounding variable: time. Now any difference between control and experimental groups cannot be attributed with certainty to the variable of interest because it could simply be that conditions have changed over time. It is very difficult to persuade the Devil's Advocate that the conditions, the equipment, or the scientist carrying out the procedures have not changed in some critical way between the time of measurement of the experimental subjects and the time of the controls. This problem is very similar to the problems of historical controls. Our colleague would have been well advised to avoid this situation by running experimental groups and controls concurrently and by randomizing the order in which procedures are applied to experimental subjects (see Sections 7.2.2 and 11.1 for more on this).

Q 2.5

The first thing to establish is whether you can find a study area that has a good number of geese flocks. Geese are sensitive to human disturbance so you need to explore how close you can get to a flock without disturbing it; how close you can get is going to influence the quality of measurements you can take on each flock. Let's keep things simple and assume that only you are collecting data—if this were not the case then the pilot study would allow you to explore the extent of between-observer variation. The pilot study would allow you to explore how well you can identify geese to species under field conditions. It might alert you to whether mixed-species flocks occur and allow you to develop rules on how to treat these. Geese flocks can feature over a hundred moving individuals, so counting them seems problematic. A pilot study would let you explore how confident you feel about estimation of flock size. One solution to counting might be to take a photograph of the whole flock and count individuals on the photograph; the pilot study would let you explore how often the terrain allows you to find a vantage point from which a suitable photograph can be taken. Prior to starting the study you will have decided on some criteria for categorizing land use type, and the pilot would allow you to assess how effective you feel this is, and if necessary fine-tune it before the main study.

Chapter 3

Q 3.1

Foraging behaviour is a plausible third variable. It is plausible that different badgers have different diets (perhaps through variation in local food availability) and that some food types are associated both with a higher risk of infection with gut parasites and with lower calorific value, leading to lower body weight.

Notice that if such a third-variable effect is operating, this does not mean that intestinal parasite load has no direct effect on body weight: both the third-variable and direct effects could be in operation, but further manipulative study would be required to estimate the relative importance of the two mechanisms for body weight.

Q 3.2

(a) There is a danger of third-variable effects here. For example, it seems likely that those who feel themselves to be time-stressed might (separately) have a high propensity to use highly processed foods and also a high propensity for depression. There is also a possibility of reverse causation, where people who are depressed for whatever reason are less likely to be motivated to make meals from scratch and more likely to turn to easier-to-prepare highly processed foods. Manipulating someone's diet is reasonably straightforward, and results ought to be seen after weeks or months. So we would opt for manipulation.

(b) Again there are a lot of potential third variables here, e.g. sex, age, socio-economic group, ethnicity, other dietary factors (such as salt content), genetic factors, and exercise. Hence, you are drawn towards manipulation. But heart attacks are relatively uncommon, and so even with thousands of people in the study, you would require the study to run for several years to hope to see any effect. There is also serious ethical concern about manipulation here, in that asking a volunteer to take a diet that you expect may increase their chance of a life-threatening event like a heart attack seems very hard to justify. In contrast, with careful monitoring, an increased risk of depression in (a)—while still an ethical issue—is perhaps more justifiable. These practical and ethical concerns would turn us away from manipulation in (b) and towards adopting a correlational approach, accepting that our data collection and statistical analysis will have to be carefully designed to control for several potentially important third factors.

Chapter 4

Q 4.1

Probably not. Married people are a subset of the population, and we would be safe to generalize from a sample of married people to the whole population only if we could argue that a person's likelihood of marriage and their height are likely to be entirely unrelated. However, this would be a difficult argument to make. Height is generally considered to be a relevant factor in people's attractiveness, which again might presumably be linked to their likelihood of marriage.

Q 4.2

An unfair question would be phrased in such a way to encourage the respondent (consciously or unconsciously) to answer in a particular way. Obviously unfairly phrased questions would include:

'Do you agree with me that Pepsi tastes better than Coke?'

'Recent surveys have shown that Coke tastes better than Pepsi, do you agree?'

'Sales of Pepsi continually outstrip Coke, do you think it tastes better?'

'I represent the Coca-Cola company, do you feel that our product tastes better than Pepsi?'

You might want to be careful even with the very neutral 'Do you prefer Pepsi or Coke?' by alternating it with 'Do you prefer Coke to Pepsi?', just in case ordering has an effect on people's answers.

Q 4.3

Read the literature. Small mammal trapping is such a common scientific technique that a great deal is probably known about bait preference and 'trapability' in your study species. You would be particularly concerned for this study if there were reports of variation between the sexes in factors that might affect their probabilities of being trapped: for example general activity level, food preference, or inquisitiveness. If these effects seem to be strong, then your survey may need a drastic rethink.

It might also be a good idea to vary potentially important factors, such as trap design, bait used, and positions where the traps are set. This would let you explore statistically whether the estimate of sex ratio that you get depends on any of these factors. Strong differences in sex ratio obtained by traps with different baits, say, would suggest that sex differences in attraction to bait are likely to be a concern in your study.

Chapter 5

Q 5.1

We would be concerned that a given female's choice could be influenced by the decisions made by others, thus making individual choices no longer independent measures of preference. This might occur because individuals are more likely to adopt what appears to be the popular choice, simply because of a desire to be in a group with a large number of other (female) birds. Alternatively, we might find that aggression or simple lack of space caused some individuals to move to the side of the cage that they would not have preferred if they had had a free choice.

Q 5.2

For hair colour you would probably be safe to count each individual as an independent data point. You might worry slightly that there may be some pseudoreplication due to relatedness if people who are particularly nervous about visiting the dentist or who are facing quite extensive procedures are accompanied by a family member, but this will be sufficiently uncommon that you may be safe to ignore it. However, pseudoreplication is a real problem for conversation. Firstly, if one person is talking then the strong likelihood is that they are talking to someone else! Hence people's decisions to talk are not independent. Further, if one conversation is struck up between two people, this is likely to lower other people's inhibitions about starting up another conversation. Hence people cannot be considered as independent replicates and you would be best to consider the 'situation' as a single replicate, and record total numbers of men and women present and the numbers of each that engage in conversation.

Q 5.3

You should be concerned. While you have 360 red blood cell counts, these only come from 60 different people. The six counts taken from each individual are not independent data points for examining the effects of the drug on people's red blood cell counts.

The simplest solution is to use your six data points to generate a single data point for each person. In this case, the mean red blood cell count would probably be appropriate. While more complicated multi-level statistical procedures would allow you to use all of these data points in your analysis, even these procedures would not allow you to use the 360 measures as independent data points.

Q 5.4

The answer to this question depends on the biology of the system. In particular it will depend on how much variation there will be between the three counts of the same sample, and also between the two samples from the same individual. If we know that there is little variation between counts of the same blood sample, then a single count of each sample might be sufficient. However, if there is a great deal of variation between smears, then taking a number of measures from a single sample and generating a mean value may give us a much better estimate of the 'true' cell count for the sample than any single count would. The same logic will apply to whether we need one or several blood samples from each individual. Thus, repeated sampling in this way may be an efficient way of dealing with some of the random variation in your system.

Chapter 6

Q 6.1

The answer depends on how big the effect you are looking for will be. If it is large then the additional power of the extra animals is not justified. If it is intermediate then the larger experiment is clearly the better, and the smaller not worth carrying out. If the effect is smaller than 0.5, then neither would be worth pursuing. It feels like it might be time to consider whether you can reduce variation, or maybe measure something else.

Q 6.2

In the first case, after congratulating them on thinking so hard about their study, you might encourage them strongly to think about reducing the number of captive primates used. 96% is a very high power, and the ethical issues (as well as the cost implications) of using more than is absolutely necessary for your study are extremely serious.

In the second case, the cost and welfare implications are far less. However, measuring leaf damage takes time that could be used for other work (or sleep!), and so again, you would advise them (less strongly this time) to consider a smaller experiment.

Q 6.3

The main advantage is that it is possible to obtain large numbers of genetically identical individuals, which in principle should be less variable than genetically diverse individuals. This may increase the power of your study.

Q 6.4

The effect of increasing the sample sizes should be beneficial for power, but the addition of a second incubator will add an extra source of random variation, potentially reducing power. Unless the incubators differ hugely or the experiment is already very large, the latter is likely to be outweighed by the former. Either way, running the experiment using a blocked design (see Chapter 9) would be advisable.

Q 6.5

The potential advantage is that having a greater difference in tail lengths may increase the effect size, and so increase the power of the study. However, the results might be criticized on the grounds that the range is beyond that which females are likely to encounter, and so their behaviour is not natural.

Chapter 7

Q 7.1

We wouldn't. There is no strong ethical imperative to minimize experiment size in an experiment on tomato plants. Further, although you might have data on tomato growth rates without this plant feed from previous experiments that could function as a historical control, could you convince the Devil's Advocate that the plants were genetically similar, grown in the same medium and under exactly the same environmental conditions? This is doubtful, so a concurrent control is almost certainly the better option.

Q 7.2

Not really, because now the rate of application is confounded by time. Let us say that the control group plants all grew at the same rate in all three experiments, but that plants given the high dose of plant food grew more vigorously than those on the low dose; surely this is conclusive proof that dose rate has an effect? No, it could simply be that application of plant food (regardless of the dose) had a stronger effect on growth under the environmental conditions that prevailed in our high-rate experiment than under those that occurred during our low-rate experiment. This might seem implausible to you, but don't expect the Devil's Advocate to be easily persuaded. This problem goes away if you carry out the treatments concurrently.

Q 7.3

Disease would be a worry, so we would thoroughly clean the greenhouse and remove unnecessary clutter before we start. We would use a fresh batch of growth medium, new pots, and sterilized equipment. We would preferably use a greenhouse where we can minimize human traffic near our experimental plants, and we could add signage that explains that these plants are part of an ongoing experiment, and gives our contact details (asking anyone to contact us if they have to move the plants or work in close proximity to them—no matter how briefly—for any reason). However, the longer we run the experiment the higher the risk of drop-outs. But here there is a trade-off with statistical power, because the longer we run the experiment the more obvious any difference in growth rates between groups will be. Hence your decision about how long to run the experiment will reflect what you feel is the best compromise between these competing factors.

Q 7.4

There is a problem here in that the beans have not been randomly allocated to treatments. Instead, the beans have been allocated to treatments based on the decisions of the female beetles as to how many eggs to lay on each bean. Beans will differ in many ways, some of which may affect beetle larva development. Unless we randomly allocate beans to the two treatments, this may lead to systematic bias. So imagine that females assess the quality of a bean and then lay a number of eggs that reflects the quality. This would mean that beans with two eggs would be, on average, better quality than beans with a single egg, and so our competition treatment would be confounded with bean quality. A better approach would be to choose beans that females had laid two eggs on, randomly allocate them to treatments, and then remove one egg from the beans in the no-competition treatment group.

Chapter 8

Q 8.1

It is tempting to assume that what this means is that the effect of one factor is affected by the values of the other factors, but actually the situation is more complicated than this. That would be the explanation if there were several interactions between just two of the factors. The interaction between three factors means something else again. The closest we can get to this is to say that not only does the effect of insecticide treatment depend on the level of food supplied (i.e. there is an interaction between insecticide and food), but that this interaction itself depends on the variety of tomato plant we are talking about. One way that this might happen is if food and insecticide affect growth rate independently in one of the varieties, but in another variety adding insecticide has a much greater effect in high-food treatments than in low-food treatments. Put another way, it means that it is only sensible to look for interactions between two factors (e.g. food and insecticide) for each specific value of the other factor (each variety of plant separately).

Q 8.2

We would worry about this conclusion for a number of reasons. First, we would be concerned that the two experiments were carried out at different times, and so any differences between the experiments might be due to tin, but might be due to anything else that has changed in the lab between experiments. We would also want to know where the plants had come from, to reassure ourselves that they were really independent replicates. However, our main concern here is about interactions. The hypothesis is about an interaction. If the plants have evolved to be able to deal better with tin, we might predict that mine plants will grow better than other plants in the presence of tin, and the same or worse in the absence of tin. In other words, the effect of one factor (plant origin) depends on the level of the other factor (tin presence or absence). By carrying out two experiments and analysing them separately, the professor has not tested for an interaction, and so his conclusions are premature.

Q 8.3

You'll need to repeat the experiment several times. Our advice would be to run the experiment using all three incubators three separate times. At each time, each temperature is assigned to a different incubator, but the assignment changes between times such that each incubator is used to produce each of the three temperatures over the course of the three experimental times. You would then be able to test for any effects of incubator and time of experiment on growth rates in your statistical analysis. More usefully, your analysis will be able to control for any effects of these factors, making it easier for you to detect effects due to the factors that you are interested in (temperature and growth medium).

Chapter 9

Q 9.1

Possibly. Blocking is only a good idea if you know that your blocking variable actually explains a substantial fraction of variation: that is, the two scientists do really score differently from each other. We'd recommend a pilot study where each scientist scores the same selection of tomatoes independently of the other, and they then compare their scores to see if there is discrepancy between them. If there is discrepancy, they should try to understand their disagreements and work towards agreeing, then repeat the trial on another set of tomatoes to see if they are now in agreement. If they now agree perfectly, then we can stop worrying about variation due to the identity of the scorer. If they still cannot agree, then it is worth considering if it is possible for one person to do all the scoring and for them to agree on who that person should be. The moral of all this is that it is better to remove the source of the unwanted variation than to use blocking or other measures to control for it.

However, if both scientists are required and they cannot be relied upon to score the same tomato the same way, then blocking on identity of the scorer would be a way to control for this unwanted variation. Each scorer should sample half of the plants in each greenhouse and half of the plants in each treatment group. Be careful when allocating plants to scientists that you don't do this systematically by having a rule like 'Scientist A does the half of the greenhouse nearest the door'. If you do this and find an effect of scorer, you will not know if it is an effect of scorer or an effect of placement within greenhouses. Allocating plants randomly to scorers is going to be a non-trivial amount of work, so you may well want to rethink looking for ways to reduce between-observer variation rather than controlling for it (see Chapter 11 for further discussion on this).

Q 9.2

The subjects are likely to be quite variable in their initial level of fitness, so blocking seems attractive. Although age and sex are easy to collect data on, we're not sure that we expect these to be strong contributors to between-subject variance in fitness. You'd be better off blocking according to some measure of initial fitness at the start of the experiment. You might ask volunteers to fill in a questionnaire about their existing level of exercise, or you might take measurements on them to calculate some index of body fat.

Chapter 10

Q 10.1

It is likely that the farm keeps records of eggs collected from different henhouses. If so, it would be worth looking at these records to see if some houses consistently produce more eggs than others. If they do, then it would be worth blocking on previous egg production.

Q 10.2

Yes it would; however, this design will not be much (if any) easier or cheaper to perform than the counterbalanced one that we recommend, but is likely to be much less powerful. One reason for this is that we will have to make between-group comparisons, and the other reason is that we will now have only half the number of within-subject comparisons to detect an effect of the experimental treatment.

Chapter 11

Q 11.1

You should record measurement date for each chick and include it as a covariate in your statistical analysis. However, you may also want to explore whether the person doing the measuring is drifting over the course of the experiments: we suggest that you get another couple of properly trained measurers to take measurements of some of the same chicks as your main measurer: this should be done entirely independently of each other and the measurements taken by the main measurer. Further, the main measurer should be told that some of their measurements will be checked independently but not which ones. Comparison between the three sets of measurements can then highlight any drift in the main measurer, since it is very unlikely that the three measurers will drift in the same way. Further, human nature being what it is, knowing that some of their measurements will be checked independently will make the main measurer particularly careful, making drift less likely in the first place. Finally, you should always bear in mind that, despite your best efforts, you may not be able to entirely remove or control for any effect of time in this study, and should consider what effects this might have on the results you have collected and the conclusions you can draw.

Q 11.2

In this case, we think not. Using two people with two stopwatches, some volunteers, and some care, you can easily collect data in a pilot study that should show that both imprecision and inaccuracy are not going to impact strongly on the results of your main experiment. Further, the cost of two stopwatches is going to be less than 1% of the cost of the automated system. It may involve increased labour costs, but labour is cheap (or cost-free if you have friends). The set-up time of coordinating the stopwatch operators, practising working together, and collecting the pilot data will probably be less than 10% of the set-up time of the automated system.

We might reach a different conclusion if the times that were being recorded were much smaller, such that the relative importance of imprecision introduced by the stopwatches increased. Thus, if we were recording speeds of racing cars over 60 metres, then the extra hassle of the automated system would be worthwhile.

Q 11.3

Many aspects of the sites could affect beetle diversity. As well as your interest in tree type, such variables as altitude and land area covered could well have an effect. If we have only two forests, they are likely to differ in other ways as well as in tree type, for example differing in altitude. With only two forests we have no way of knowing whether any difference in beetle diversity found in the two sites is due to the variable of interest (tree type) or some third variable (for example, altitude). We need more samples so we can control for the effects of these potential third variables statistically. Another way to think of this is that the measurements of multiple trees in a single forest are not independent measures of that kind of forest, and so they are pseudoreplicates if we want to address questions about differences between types of forest.

Q 11.4

We cannot take the approach recommended for the satellite images, because we cannot measure the same cats on more than one occasion. Even if cats could be brought in more than once, they may well have experienced weight change over the intervening time.

You must minimize the room for inadvertent variation in measurement technique. Have a strict written protocol for the measurement. This should include the posture of the cat while you are measuring the length of its back, with clear rules for identification of the two endpoints (top of the head and base of the tail). The mass should be taken on an electronic balance, calibrated at regular intervals. We expect greater scope for error in the length measurement than the mass measurement: hence we suggest that you measure the length at the start, then weigh the cat, before finally taking the length measurement again, giving you two estimates of the harder-to-measure variable on each cat.

We further suggest that you organize for a third party to repeat some of your measurements without your knowledge, in a similar manner to as discussed for Q 11.1.

Lastly, if there are likely to be concerns about observer drift over time, then it is imperative that order of measurement is randomized across cats on different treatments if at all possible.

Q 11.5

You are introducing an unintended confounding variable: identity of observer. If you find a difference between two treatment groups you do not know if this can be explained in whole or in part by differences in the way the two observers recorded their data.

Q 11.6

Firstly, we think you need each subject to be evaluated independently by several psychiatrists who should not have a history of close association with each other. We would also recommend that if possible you avoid using psychiatrists who have previously expressed strong views on gender issues. And as in all situations where it is impossible to have the perfect design, if you cannot remove the potential for observer bias completely, you should be clear about what effect it might have on your results and conclusions.

Q 11.7

Firstly consider whether you need to interview the subjects, or whether asking them to respond to written questions could be just as effective without the risk of observer effects.

Make sure that the interviewer is someone that the interviewees consider to be disinterested in them as a specific person. For example, they should be someone that the interviewee is unlikely to have encountered before or encounter again.

Train your interviewer very carefully not to (often subconsciously) encourage certain types of answers, and design your preparation of the subjects for interview so as to make the subjects less susceptible to observer effects (e.g. by emphasizing the anonymity of their responses).

Q 11.8

We don't know! But if your experiment involves collecting data in this way then you need to know! You need to do a pilot study where you collect data in this way for 30 minutes, spending the last 5 minutes rescoring the same slides that you did in the first 5 minutes. If you have good agreement then it seems that fatigue is not a problem over 30-minute timescales, but without the pilot study you would not have known, and would have no defence against a Devil's Advocate who expected that fatigue could easily kick in that quickly.

Flow chart on experimental design

The following flow chart is intended to take you step by step through the different stages involved in designing an experiment. You can treat it a bit like a checklist that should be run through every time you plan an experiment to make sure you've considered all the important issues that we mention in this book. Taking account of these issues properly will make the difference between a successful experiment that's going to yield valuable results, and an experiment that might give results that are all but worthless. Think of it this way: the time you will need to spend designing an experiment carefully will only ever be a fraction of the time you actually spend gathering results. Just a few extra minutes early on should mean that you'll be able to feel much more confident during the many more minutes you're going to spend in the lab, or out in the field. It may seem laborious but, trust us, it will be time well spent. Good luck!

The numbers on the chart provide direction to the relevant sections of the book, for each part of the design process.

(i) Preliminaries

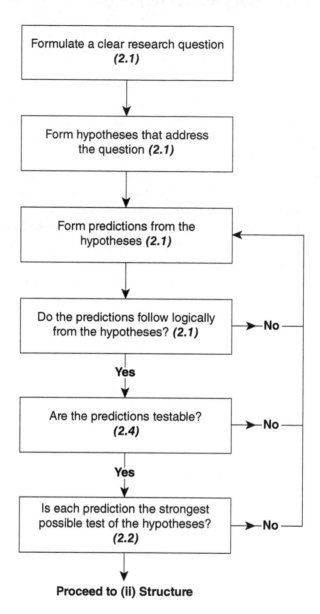

(ii) Structure

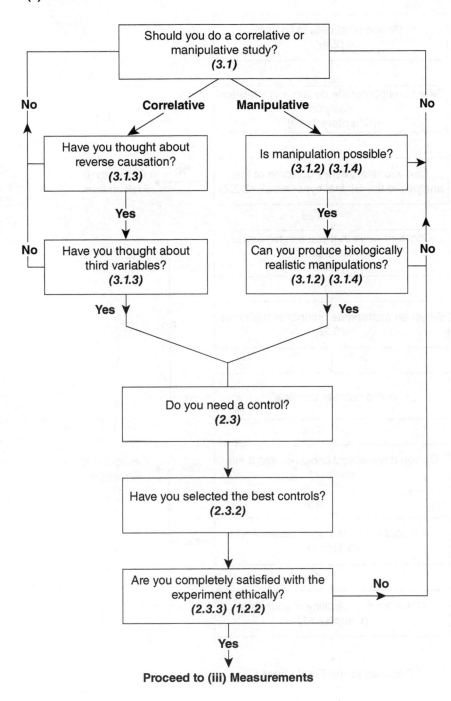

(iii) Measurements

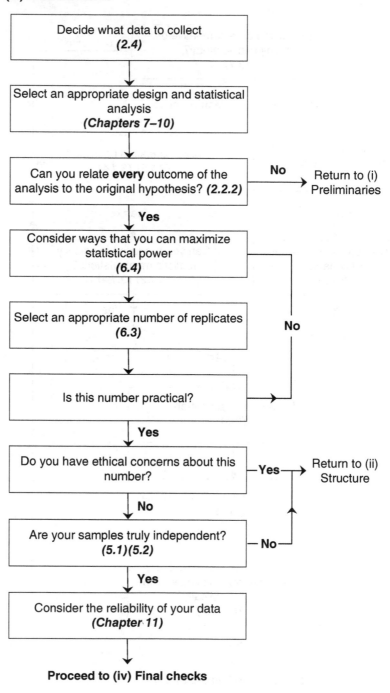

(iv) Final checks

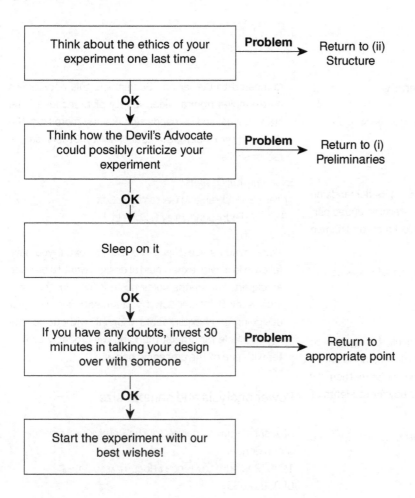

Bibliography

Further books on design generally

Principles of Experimental Design for the Life Sciences
Murray R. Selwyn
1996 CRC Press
0849394619

Very clearly written and very gentle in its demands on the reader's mathematical abilities. Recommended particularly for those interested in clinical trials and human medicine generally.

Experimental Design and Analysis in Animal Sciences
T.R. Morris
1999 CABI Publishing
0851993494

A gentle first step beyond our book, taking you to places that we don't go, such as towards unbalanced designs. Makes little demand of the reader mathematically. As the title suggests, a good buy for students of Agricultural Sciences.

Practical Statistics and Experimental Design for Plant and Crop Science
Alan G. Clewer and David Scarisbrick
2001 John Wiley & Sons Ltd
047899097

This is to be strongly recommended, whether you are a crop scientist or not. It is a very good introduction to statistics and links this very well with experimental design. It asks a little more of the reader in terms of mathematical ability, but is a model of clarity and interesting writing. Many scientists will never need to know more about experimental design and statistical analysis than is contained in this book.

Introduction to the Design and Analysis of Experiments
Geoffrey M. Clarke and Robert E. Kempson
1997 Arnold
978-0340 645555

Compared to Clewer and Scarisbrick, this book draws on examples from a wider variety of branches of the life sciences, and demands significantly more from the reader in terms of mathematical ability. Otherwise it is rather similar.

Experimental Designs
William G. Cochran and Gertrude M. Cox
1957 (2nd edn) John Wiley & Sons Ltd
0471545678

Never mind its age, this is the classic text. If you have fallen in love with experimental design, want to become an expert, and maths holds no fear for you, then you must seek this book out. If you are satisfied with just designing good experiments yourself, without becoming an oracle for others, then you will probably be better with one of the preceding books.

Power analysis and sample size

Statistical Power Analysis for the Behavioural Sciences
Jacob Cohen
1988 (2nd edn) Lawrence Erlbaum Associates
0805802835

Don't be put off by it being a few years old; this is the definitive text. The only drawback is that it is about four times the price of the two alternatives that we list.

Design Sensitivity: Statistical Power for Experimental Research
Mark W. Lipsey
1990, Sage Publications Inc (USA)
0803930631

Aimed at social scientists, but very clearly written and very much in a similar vein to Cohen. So if you are unsure about taking the plunge, then this might be the best way in. Either it will satisfy you, or it will make your eventual jump to Cohen a little less daunting.

Statistical Power Analysis
Kevin R. Murphy, Brett Myors, and Allen Wolach
2014 Routledge
1848725884

By far the least intimidating of the three books we recommend. It's on our approved list because it is very clearly written, and whilst some ability to handle equations is required to make much use of the other two, this is written in such a way that those with very little confidence in mathematics could make some use of it. You could read it from start to finish in a day, but for all that, it is not insubstantial, and for some readers will provide all the power analysis they need.

In very large-scale ecological experiments, for example, where an entire lake or river system must be manipulated, then replication may not be feasible. However, before arguing that you don't need replication in such experiments, think carefully. To help in this, we recommend reading Oksanen, L. (2001) Logic of experiments in ecology: is pseudoreplication a pseudoissue? *Oikos* **94**, 27–38.

There are a lot of software packages (many available free on the Internet) that will do power calculations for you. Simply type 'statistical power calculations' into your favourite search engine. A newly updated version of an old favourite, G*Power is available free from http://www.psycho.uni-duesseldorf.de/abteilungen/aap/gpower3/.

For more thoughts on using unbalanced designs see Ruxton, G.D. (1998) Experimental design: minimizing suffering may not always mean minimizing number of subjects. *Animal Behaviour* **56**, 511–12.

Randomization

The paper that brought the idea of pseudoreplication to a wide audience was that of Hurlbert, S.H. (1984) Pseudoreplication and the design of ecological field experiments. *Ecological Monographs* **54**, 187–211.

This paper can be recommended as a good read with lots to say on design generally as well as a very clear discussion on pseudoreplication. All that and it discusses human sacrifices and presents an interesting historical insight into the lives of famous statisti-

cians. Highly recommended: over 2000 citations in the scientific literature must also be some kind of recommendation!

For further reading on the techniques required for effective comparisons between species, the following paper provides an intuitive introduction: Harvey, P.H. and Purvis, A. (1991) Comparative methods for explaining adaptations. *Nature* **351**, 619–24.

If this seems of interest to you, then you might want to look at one of the following books

The Phylogenetic Handbook Second Edition:
A Practical Approach to Phylogenetic Analysis and
Hypothesis Testing
Edited by Phillippe Lemey
2009 Cambridge University Press
0521730716

Modern Phylogenetic Comparative Methods and Their
Application in Evolutionary Biology:
Concepts and Practice
Edited by Laslo Zsolt Garamszegi
2014 Springer
3662435497

Your first book on statistics

Choosing and Using Statistics
Calvin Dytham
2010 (3rd edition) Wiley Blackwell
1405198397

Those who like this book absolutely love it. Its great strength is that it can tell you very quickly what statistical test you want to do with your data and how to actually do that test on numerous different computer packages. This is a book about using statistics, not about the philosophy of statistical thinking.

Practical Statistics for Field Biology
Jim Fowler, Lou Cohen, and Phil Jarvis
1998 John Wiley & Sons Ltd
0471982962

The opposite extreme to Dytham's book. There is no discussion on the details of specific computer packages and plenty on the underlying logic behind different statistical approaches. Don't get the impression that this is an esoteric book for maths-heads though—it is full of common sense advice and relies more on verbal reasoning than mathematics. Packed with illuminating and interesting examples.

Practical Statistics for Experimental Biologists
Alastair C. Wardlaw
2000 John Wiley & Sons Ltd
0471988227

Very similar to Fowler and friends. The key differences are that it gives extensive examples developed for a specific computer package (Minitab) and the examples come more from the laboratory than the field.

Biomeasurement
Dawn Hawkins
2014 (3rd edn) Oxford University Press
0199650446

Covers very similar ground to the preceding books and provides practical guidance in actually doing statistical tests in SPSS. Of all the books, this one is probably the most sympathetic to students who feel particularly nervous about statistics. However, this should not be taken as an indication that the material is dumbed down. Indeed a particular strength of this book is that it repeatedly makes use of real examples from the primary literature. Another strength is a particularly comprehensive companion website.

Asking Questions in Biology
Chris Barnard, Francis Gilbert, and Peter McGregor.
2011 (4th edn) Benjamin Cummings
0273734687

This is a lot more than an elementary statistics textbook. Rather, it looks to take students through all the various processes of scientific enquiry: from asking the right question, to turning the question into a well-designed experiment, analysing the data from that experiment, and finally communicating the outcome of the work. In 256 pages, this book cannot attempt to be comprehensive in any of the aspects it covers: but for those just starting to do their own experiments, this book is full of practical advice and wisdom.

Modern Statistics for the Life Sciences
Alan Grafen and Rosie Hails
2002 Oxford University Press
0199252319

The word modern in the title should alert you that this book takes a rather different approach to traditional elementary statistical texts. Rather than introduce a number of statistical tests as discrete tools, it suggests an overarching framework based on construction of statistical models and the powerful and flexible modern technique of the General Linear Model. Our guess is that some day most statistics courses will be taught this way; in the meantime the traditional method still dominates. This book will appeal more to the student who wants to develop intuitive insight into the underlying logic of statistical testing.

More advanced (and expensive) general statistical books

Biometry
Robert R. Sokal and F. James Rohlf
2012 (4th edn) W.H. Freeman
0716786044

For the issues it covers, Sokal and Rohlf is definitive. It does not cut corners, shirk from difficult issues, or encourage you to be anything other than very, very careful in your analysis. If your experiment has features that make your situation a little non-standard, and your elementary textbook doesn't cover the problem, then almost certainly these guys will. It's not an easy read, but if you put in the time it can deliver solutions that you'll struggle to find in easier books.

Biometrical Analysis
J.H. Zar
2013 (5th edn) Prentice Hall
1292024046

Some find Sokal and Rohlf a little stern; if so, Zar is the answer. Compared to elementary books, there is a lot more maths, but the book does take its time to explain clearly as it goes along. Highly recommended.

Experimental Design and Data Analysis for Biologists
Gerry P. Quinn and Michael J. Keough
2002 Cambridge University Press
0521009766

Less exhaustive in its treatment than either of the above, but this allows it to cover ground that they do not, including a considerable treatment of multivariate statistics. It also covers more experimental design than the other two. It has a philosophical tone that encourages reading a large chunk at one sitting. However, it's also a pretty handy book in which to quickly look up a formula. This should be a real competitor for Zar. Try and have a look at both before deciding whose style you like best.

Designing Experiments and Analyzing Data: A Model Comparison Approach
S.E. Maxwell and H.D. Delaney
2003 (2nd edn) Taylor and Francis
978-0805837186

If you really want to get to grips with the link between experimental design and statistical models, this is an excellent place to look. Explains clearly how statistical model fitting works, in a way that allows the reader to understand as well as carry out the analyses themselves.

Experiments in Ecology
A.J. Underwood
1996 Cambridge University Press
0521556961

A blend of design and statistics, presented in an unusual way that will be perfect for some but not for all. Look at this book particularly if ANOVA is the key statistical method you are using. Good on subsampling too.

Within-subject designs

All of the more advanced statistics texts just discussed with cover within-subject designs, but if you really want to immerse yourself in this subject then try the definitive text:

Design and Analysis of Cross-Over Trials
B. Jones and M.G. Kenward
2014 (3rd edition) Chapman & Hall/CRC
1439861420

Also excellent is

Cross-over Trials Clinical Research
S. Senn
2002 (2nd edition) Wiley
0471496537

Ecology, ethology, and animal behaviour

Collecting data can be very challenging in these fields. Fortunately, there are some wonderful books to help you do this well.

Measuring Behaviour: An Introductory Guide
Paul Martin and Patrick Bateson
2007 (3rd edn) Cambridge University Press
978-0521535632

Many professors will have a well-thumbed copy of a previous edition of this on their shelves, which they bought as an undergraduate, but still find the need to consult. Brief, clear, and packed with good advice—cannot be recommended highly enough.

Observing Animal Behaviour
Marian Stamp Dawkins
2007 Oxford University Press
019856936X

A very short book but packed with common-sense practical advice and one of the most thoughtful books on helping you think effectively as a scientist.

Ecological Census Techniques
William J. Sutherland
2006 (2nd edition) Cambridge University Press
0521606365

Quite simply, if you are going to be censusing a population of any wild organism this book should be your first port of call.

Ecological Methodology
Charles J. Krebs
1999 (2nd edn) Addison-Welsey
0321021738

Much more substantial, and positively encyclopaedic on the design and analysis of ecological data collection. Beautifully written and presented; often funny and very wise.

Handbook of Ethological Methods
Philip N. Lehmer
1996 (2nd edn) Cambridge University Press
0521637503

Similar in many ways to the Krebs book in what it seeks to cover. Has a stronger emphasis on the mechanics of different pieces of equipment and good advice on fieldcraft, but less emphasis on analysis of the data after it is collected.

Index